International Conference on Mechanics and Applied Mathematics (MAM 2022)

Journal of Physics: Conference Series Volume 2455

Ghaungzhou, China
18-20 November 2022

ISBN: 978-1-7138-7166-8
ISSN: 1742-6588

Printed from e-media with permission by:

Curran Associates, Inc.
57 Morehouse Lane
Red Hook, NY 12571

Some format issues inherent in the e-media version may also appear in this print version.

Additional copies of this publication are available from:

Curran Associates, Inc.
57 Morehouse Lane
Red Hook, NY 12571 USA
Phone: 845-758-0400
Fax: 845-758-2633
Email: curran@proceedings.com
Web: www.proceedings.com

TABLE OF CONTENTS

APPLIED MECHANICS AND MECHANICAL MODEL ANALYSIS

MATHEMATICAL MODELING AND GEOMETRIC EQUATION CALCULATION

Preface

The 2022 International Conference on Mechanics and Applied Mathematics (MAM 2022) was successfully held in Guangzhou, China from November 18th-20th, 2022 (virtual conference). The conference is a leading international academic meeting and has the potential to bring new horizons and insights to the researchers and practitioners in the fields of mechanics and applied mathematics.

The close relationship and interaction between mechanics and mathematics and the research activities on them has promoted the stable development of society and economy. MAM 2022 created an academic platform for scientific innovation, brought together experts, scholars, and scientists from all over the world to present their work and exchange information, promoted the industrial cooperation of academic achievements, and facilitated future business contacts among all the participants.

The scope of the conference covered all areas of research in mechanics and applied mathematics, including Newtonian Mechanics, Explosion Mechanics, Elementary Algebraic Topology, Discrete Mathematics, Coupled Numerical Simulation, Mathematical Methods in Mechanics, etc. The conference brought together about 100 participants, primarily from China and India, but also from Malaysia, French, and other countries.

A three-day program of the conference was divided into three sessions, including keynote speeches, oral presentations, and poster presentations. Each keynote speaker was given 30-40 minutes for keynote speeches and every presenter was given 10-15 minutes. Among them, Prof. Dr. Antti Rasila from Guangdong Technion-Israel Institute of Technology, China performed a keynote speech on the title: Conjugate Function Method and Numerical Conformal Mappings. He discussed the so-called conjugate function method, an algorithm based on properties of so-called conformal invariants, to obtain numerical representations of conformal mappings and its applications. Besides, he presented the basics of the algorithm, and discussed its potential applications related to various engineering problems, for example, stress analysis, crack propagation, inverse problems, and mesh generation. The insightful speeches of each keynote speakers triggered interesting discussions and a number of informal talks between all the participants.

We would like to thank all the participants in the 2022 International Conference on Mechanics and Applied Mathematics. We would also like to thank the Organizing and Technical Program Committee, for the work they put in to organize this event. Finally, we would like to acknowledge the Journal of Physics: Conference Series, for the endeavor of all its colleagues in publishing this paper volume. We deeply appreciate their efforts in making the conference successful.

The Committee of MAM 2022

Committee member

Conference General Chair
Prof. Rozaini Bin Roslan, Universiti Tun Hussein Onn Malaysia, Malaysia

Publication Chair
Prof. Dragana Cvetković Ilić, University of Niš, Serbia

Technical Program Committee Chair
Prof. Erdal Karapinar, Cankaya University, Turkey

Technical Program Committee
Prof. Hakan ÇALIŞKAN, USAK UNIVERSITY, Turkey
Prof. Hakim NACEUR, INSA Hauts-de-France, French
Prof. Koustubh Ajit Supriya Kabe, Mumbai University, India
Prof. MEGHA M. UPLANE, Savitribai Phule Pune University, India
Prof. Mohamed Abdel Kader El Gelany Ismail, Miami College of Henan University, Canada
Prof. Xingli Li, Taiyuan University of Science and Technology, China
A. Prof. Md Sazzad Hossien Chowdhury, International Islamic University Malaysia, Malaysia
A. Prof. RIZWAN UL HAQ, Bahria University, Pakistan
A. Prof. Srimanti Roychoudhury, Adamas University, India
A. Prof. Akhlaq Husain, BML Munjal University Gurgaon, India
A. Prof. Mohammad A. Gharaibeh, The Hashemite University, Jordan
A. Prof. Mohammad Russel, Dalian University of Technology, Bangladesh
A. Prof. Md Sazzad Hossien Chowdhury, International Islamic University Malaysia, Bangladesh

Organizing Committee
Assistant Professor, Camelia Enzevaee, Al-Zahra University, Iran
Assistant Professor, Ali Rahman, Southwest Jiaotong University, China
Assistant Professor, Alfrendo Satyanaga Nio, Nazarbayev University, Singapore
Dr. Shubham Sharma, IK Gujral Punjab Technical University, India
Dr. Rameeza Moideen, Indian Institute of Technology Bombay, India
Dr. Mohammad Fadzli bin Ramli, Universiti Malaysia Perlis, Malaysia
Dr. Fauziah Marpani, Universiti Teknologi Mara, Malaysia
Dr. Vasos Pavlika, University of London, UK
Dr. Navid Bayati, Aalborg University, Danmark
Dr. Morteza Hosseini, Eindhoven University of Technology, Holland
Dr. Nor Suriya binti Abd Karim, Universiti Pendidikan Sultan Idris, Sudan
Dr. Sangeeta Verma, National Institute of Hydrology, India

Peer Review Statement

All papers published in this volume have been reviewed through processes administered by the Editors. Reviews were conducted by expert referees to the professional and scientific standards expected of a proceedings journal published by IOP Publishing.

- **Type of peer review:** Single Anonymous
- **Conference submission management system:** Morressier
- **Number of submissions received:** 69
- **Number of submissions sent for review:** 61
- **Number of submissions accepted:** 32
- **Acceptance Rate (Submissions Accepted / Submissions Received × 100):** 46.4
- **Average number of reviews per paper:** 2
- **Total number of reviewers involved:** 12
- **Contact person for queries:**
  Name: Xuexia Ye
  Email: xx.ye@keoaeic.org
  Affiliation: AEIC Academic Exchange Information Centre

Dynamic Buckling Analysis of the Cylindrical Shell with Radial Stiffness and Damping Constraints under Axially Impact

Yifei Gui[1], Jiaxin Xu[2], and Juhui Ren[*3]

[1] School of Mechanical Engineering, Shanghai DianJi University, Shanghai, 201306, China

[2] Shanghai Aircraft Design and Research Institute, Commercial Aircraft Corporation of China, Shanghai, 201210, China

[3] School of Electronic Information Engineering, Shanghai DianJi University, Shanghai, 201306, China

[*]Corresponding author e-mail: renjh@sdju.edu.cn

Abstract. The governing equation of a cylindrical shell with radial stiffness and damping constraints is established and then solved. The bifurcation conditions of buckling are obtained given boundary conditions and consistency conditions. The critical loads and buckling modes are analyzed. The results reveal both radial stiffness and damping constraints will enhance the anti-buckling capacity of cylindrical shells, and the capacity of the former is better than that of the latter. The radial stiffness and damping values will have a great influence on the critical load when they are small but do not have much influence on the critical load with increasing their values.

1. Introduction

The cylindrical shell is an important load-bearing structure, and its dynamic buckling under impact has always been one of the hottest issues. Much research has been taken on the buckling about a cylindrical shell on the axial impact. Ali et al. investigated dynamic nonlinear buckling about the functionally exponentially graded toroidal cylindrical shells segments, which under constant loading rates, based on the shear deformation hypothesis considering damping effect, and various parameter effects, for instance , velocity, geometrical parameters, imperfections, nonhomogeneous parameters and damping ratios, on the buckling were researched at length. Shiomitsu et al. discussed the collapse behavior and ultimate strength for ring-stiffened shells subjected to external pressure accompany with torsional buckling or local buckling of the stiffeners.

For the dynamic buckling of shells under the impact on elastic foundation, many investigations have been performed. Yang et al. presented a kind of approach about the free vibration of cylindrical shells with elastic foundations, besides, gifted any type of classical boundary conditions, some deep-seated results were presented to manifest the significant influence of foundation coefficients and geometric properties on the natural frequencies of the shell contacted the elastic medium. Srinivas et al. [1] explore the free vibration research of buried/embedded shells, these buried shells were modeled as circular shells surrounded by an elastic foundation, and the solution to the vibration problem was done by using the wave propagation method.

For the stability of constrained damping layer cylindrical shells, some studies have been reported. The emphasis of Ji et al. [2] was placed on the vibration property of a cantilever shell with MCLD treatment for the first torsional mode (1, 1), and the desirable effects about patch width and length, were analyzed, obtained results indicated that the MCLD patches at the free-end can significantly enhance system damping capacity. Jin et al. [3] discussed the applied different thicknesses about constrained passive damping layer structure vibration characteristics of short column shell through changing passive constrained damping layer structure about the thickness of viscoelastic damping constraints and layer.

Cylindrical shells have been applied extensively in engineering practice. And there are cylindrical shells embedded in the foundation in the field of civil construction, frequently. To describe its environmental characteristics accurately, a cylindrical shell model with radial stiffness and damping constraints is proposed. The buckling problem about cylindrical shells with radial stiffness as well as damping constraints is discussed by adjusting the radial stiffness and damping values.

2. Solution of governing equation

Let's conceive an cylindrical shell,which of neutral surface radius r, thickness h and length L, and constrains radial stiffness and damping. The structural material properties include Poisson's ratio μ and Young's modulus E. The shell is shown in Figure 1 (k is the radial stiffness and c_z is the radial damping).

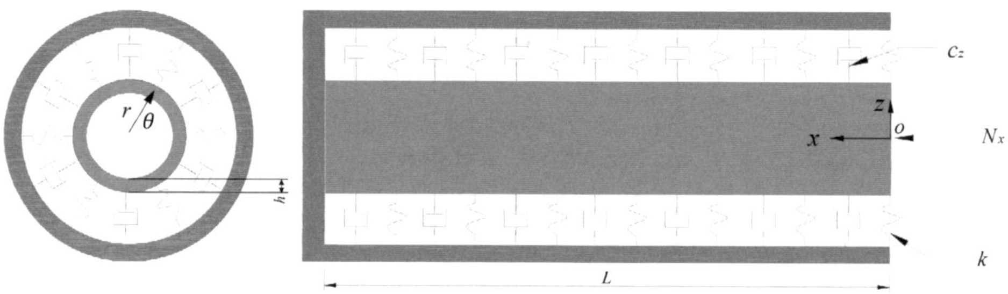

Figure 1. A cylindrical shell with radial stiffness and damping constraints

The study is settled in cylindrical coordinate system (x,θ,z). The corresponding displacements in the axial, circumferential, and radial directions are denoted by u, v, w, respectively. P_w is the radial force per unit length along the axis, and $P_w = c_z \partial w / \partial t + kw$. The governing equation of the shell is obtained by the Hamilton variation principle as follows [4, 5]:

$$D\frac{\partial^4 w}{\partial x^4} + N_x \frac{\partial^2 w}{\partial x^2} + \frac{Eh}{r^2} w + \frac{D}{r^4}\frac{\partial^4 w}{\partial \theta^4} + \frac{2D}{r^2}\frac{\partial^4 w}{\partial x^2 \partial \theta^2} + P_w = 0 \tag{1}$$

Where $D = Eh^3 / \left[12\left(1-\mu^2\right)\right]$, and N_x is the axial force borne by the cylindrical shell. The stress wave propagates in the shell in the form of a longitudinal wave when the shell is impacted by an axial step load. In the stress wave propagation process, the position about the wavefront is x_e, the position of the reflected wavefront is x_r, and the wave velocity is v_e. The axial force can be expressed as follows:

$$\begin{cases} N & 0 \le t < L/v_e, \quad 0 \le x \le x_e \\ 0 & \qquad\qquad\qquad x_e < x \le L \\ N & L/v_e \le t \le 2L/v_e, \quad 0 \le x \le x_r \\ 2N & \qquad\qquad\qquad x_r < x \le L \end{cases} \tag{2}$$

For generality and brevity, the following dimensionless variables and parameters are adopted: $W = w/r$, $X = x/r$, $T = v_e t/r$, $P = N_x/Eh$, $\xi^2 = h^2/3r^2(1-\mu^2)$, $\bar{C} = c_z v_e/Eh$, $K = 1 + kr^2/Eh$. In dimensionless terms, Equation (1) is

$$\frac{\xi^2}{4K}\frac{\partial^4 W}{\partial X^4} + \frac{P}{K}\frac{\partial^2 W}{\partial X^2} + W + \frac{\xi^2}{4K}\frac{\partial^4 W}{\partial \theta^4} + \frac{\xi^2}{2K}\frac{\partial^4 W}{\partial X^2 \partial \theta^2} + \frac{\bar{C}}{K}\frac{\partial W}{\partial T} = 0 \tag{3}$$

Let

$$W(X,\theta,T) = \Phi(X)\Psi(\theta)\Gamma(T) \tag{4}$$

substituting Equation (4) into Equation (3), the equation is obtained

$$\frac{\xi^2}{4K}\frac{\partial^4 \Phi}{\partial X^4}\Psi\Gamma + \frac{P}{K}\frac{\partial^2 \Phi}{\partial X^2}\Psi\Gamma + \Phi\Psi\Gamma + \frac{\xi^2}{4K}\Phi\frac{\partial^4 \Psi}{\partial \theta^4}\Gamma + \frac{\xi^2}{2K}\frac{\partial^2 \Phi}{\partial X^2}\frac{\partial^2 \Psi}{\partial \theta^2}\Gamma + \frac{\bar{C}}{K}\Phi\Psi\frac{\partial \Gamma}{\partial T} = 0 \tag{5}$$

Because the function $\Psi(\theta)$ periodicity in boundary θ, therefore, the boundary conditions are

$$\Psi(0) = \Psi(2\pi);\ \frac{\partial}{\partial \theta}\Psi(0) = \frac{\partial}{\partial \theta}\Psi(2\pi);\ \frac{\partial^2}{\partial \theta^2}\Psi(0) = \frac{\partial^2}{\partial \theta^2}\Psi(2\pi);\ \frac{\partial^3}{\partial \theta^3}\Psi(0) = \frac{\partial^3}{\partial \theta^3}\Psi(2\pi) \tag{6}$$

Then the function $\Psi(\theta)$ can be taken in the form

$$\Psi(\theta) = q_1\cos(\eta\theta) + q_2\sin(\eta\theta) \tag{7}$$

Where $\eta = (2\sqrt{\gamma}/\xi)^{1/2}, \gamma = \xi^2 n^4/4\ (n = 0,1,2,...)$, q_1, q_2 are constant coefficients. n denotes the circumferential buckling mode. $n = 0$ means the axisymmetric mode, and the section shape of the cylindrical shell is circular. $n \neq 0\ (n = 1,2,...)$ means the non-axisymmetric mode, and the section shape is a triangle or other polygonal shape. Substituting Equation (7) into Equation (5), the equation is obtained,

$$\frac{\xi^2}{4K}\frac{\partial^4 \Phi}{\partial X^4}\Gamma + \left(\frac{P}{K} - \frac{1}{2K}\eta^2\xi^2\right)\frac{\partial^2 \Phi}{\partial X^2}\Gamma + \left(1 + \frac{\xi^2\eta^4}{4K}\right)\Phi\Gamma + \frac{\bar{C}}{K}\Phi\frac{\partial \Gamma}{\partial T} = 0 \tag{8}$$

Equation (8) can be expressed as:

$$\frac{\Gamma}{\frac{\partial \Gamma}{\partial T}} = -\frac{\frac{\bar{C}}{K}\Phi}{\frac{\xi^2}{4K}\frac{\partial^4 \Phi}{\partial X^4} + \left(\frac{P}{K} - \frac{1}{2K}\eta^2\xi^2\right)\frac{\partial^2 \Phi}{\partial X^2} + \left(1 + \frac{\xi^2\eta^4}{4K}\right)\Phi} = \frac{\bar{C}}{K}\varpi \tag{9}$$

Where the right-hand side of the Equation (9) is a constant. In the light of the two fractions at the right-hand side of Equation (9), Equation (10) is achieved:

$$\frac{\xi^2}{4K}\frac{\partial^4 \Phi}{\partial X^4} + \left(\frac{P}{K} - \frac{1}{2K}\eta^2\xi^2\right)\frac{\partial^2 \Phi}{\partial X^2} + \left(1 + \frac{\xi^2\eta^4}{4K} + \frac{1}{\varpi}\right)\Phi = 0 \tag{10}$$

Let us take the shell of length 1, the impact end $X = 0$, and the non-impact end $X = 1$, considering the dimensionless variables. The general solution of $\Phi(X)$ can be obtained by solving Equation (10).

$$\Phi(X) = \begin{cases} a_1\cos(\alpha_1 X) + a_2\sin(\alpha_1 X) + a_3\cos(\beta_1 X) + a_4\sin(\beta_1 X) & 0 \le T < 1, 0 \le X < T \\ 0 & T \le X < 1 \end{cases} \tag{11}$$

$$\Phi(X) = \begin{cases} a_1\cos(\alpha_1 X) + a_2\sin(\alpha_1 X) + a_3\cos(\beta_1 X) + a_4\sin(\beta_1 X) & 1 \le T \le 2, 0 \le X < 2 - T \\ a_5\cos(\alpha_2 X) + a_6\sin(\alpha_2 X) + a_7\cos(\beta_2 X) + a_8\sin(\beta_2 X) & 2 - T \le X < 1 \end{cases} \tag{12}$$

Where

$$\begin{cases} \alpha_1, \beta_1 = \dfrac{\sqrt{2K}}{\xi}\sqrt{(\dfrac{P}{K}-\dfrac{1}{2K}\eta^2\xi^2)\pm\sqrt{(\dfrac{P}{K}-\dfrac{1}{2K}\eta^2\xi^2)^2-\dfrac{\xi^2}{K}(1+\dfrac{\xi^2\eta^4}{4K}+\dfrac{1}{\varpi})}} \\ \alpha_2, \beta_2 = \dfrac{\sqrt{2K}}{\xi}\sqrt{(\dfrac{2P}{K}-\dfrac{1}{2K}\eta^2\xi^2)\pm\sqrt{(\dfrac{2P}{K}-\dfrac{1}{2K}\eta^2\xi^2)^2-\dfrac{\xi^2}{K}(1+\dfrac{\xi^2\eta^4}{4K}+\dfrac{1}{\varpi})}} \end{cases}$$

a_k $(k=1,2,...,8)$ denote constant coefficients. Equation (11) and Equation (12) are the general solutions for $\Phi(X)$ before and after stress wave reflection.

3. Boundary condition and consistency condition

The cylindrical shell can be divided into the stress wave disturbance region and the undisturbed region before reflection. Then the shell can be divided into the reflected wave disturbance region and the undisturbed region after reflection. $W(X,\theta,T)$ needs to satisfy the consistency condition at the wavefront and the constraint conditions.

Assuming the free boundary at the impact end ($X=0$), and the fixed constraint at the non-impact end ($X=1$), analysis can be performed by considering $W(X,\theta,T)$,

$$\frac{\partial^2}{\partial X^2}W(0,T)=0, \frac{\xi^2}{4}\frac{\partial^3}{\partial X^3}W(0,T)+P\frac{\partial}{\partial X}W(0,T)=0$$

The above formulas can be rewritten as:

$$\frac{\partial^2}{\partial X^2}\Phi(0)=0, \frac{\xi^2}{4}\frac{\partial^3}{\partial X^3}\Phi(0)+P\frac{\partial}{\partial X}\Phi(0)=0 \tag{13}$$

$$\Phi(1)=0, \frac{\partial}{\partial X}\Phi(1)=0 \tag{14}$$

Before reflection ($0\le t<L/v_e$), the wavefront needs to satisfy the consistency condition as follows:

$$\Phi_-(X_e)=0, \frac{\partial}{\partial X}\Phi_-(X_e)=0 \tag{15}$$

After reflection ($L/v_e\le t\le 2L/v_e$), the reflected wavefront needs to satisfy the condition as:

$$\begin{cases} \Phi_-(X_r)=\Phi_+(X_r), \dfrac{\partial}{\partial X}\Phi_-(X_r)=\dfrac{\partial}{\partial X}\Phi_+(X_r), \dfrac{\partial^2}{\partial X^2}\Phi_-(X_r)=\dfrac{\partial^2}{\partial X^2}\Phi_+(X_r) \\ \dfrac{\xi^2}{4}\dfrac{\partial^3}{\partial X^3}\Phi_-(X_r)+P\dfrac{\partial}{\partial X}\Phi_-(X_r)=\dfrac{\xi^2}{4}\dfrac{\partial^3}{\partial X^3}\Phi_+(X_r)+2P\dfrac{\partial}{\partial X}\Phi_+(X_r) \end{cases} \tag{16}$$

4. Bifurcation condition of the buckling process

The next objective is to analyze the bifurcation buckling of the shell constrained by stiffness and damping. Before reflection, the boundary condition Equation (13) and the consistency condition Equation (15) need to be substituted into Equation (11), obtaining

$$\boldsymbol{A}_1\boldsymbol{C}_1=\boldsymbol{0} \tag{17}$$

Where

$$\boldsymbol{A}_1 = \begin{bmatrix} \alpha_1^2 & 0 & \beta_1^2 & 0 \\ 0 & \alpha_1(-\frac{\xi^2\alpha_1^2}{4}+P) & 0 & \beta_1(-\frac{\xi^2\beta_1^2}{4}+P) \\ \cos(\alpha_1 T) & \sin(\alpha_1 T) & \cos(\beta_1 T) & \sin(\beta_1 T) \\ -\alpha_1\sin(\alpha_1 T) & \alpha_1\cos(\alpha_1 T) & -\beta_1\sin(\beta_1 T) & \beta_1\cos(\beta_1 T) \end{bmatrix} \tag{18}$$

$$\boldsymbol{C}_1 = (a_1, a_2, a_3, a_4)^{\mathrm{T}} \tag{19}$$

If $\boldsymbol{C}_1 = \boldsymbol{0}$, then $\Phi(X) = 0$ according to Equation (11), i.e., $W(X,\theta,T) = \Phi(X)\Psi(\theta)\Gamma(T) = 0$; and buckling will not occur in this case. Large radial deformation will be generated when buckling, i.e., $W(X,\theta,T) \neq 0$. Thus, the non-zero solution for Equation (17), namely, the bifurcation condition, is

$$|\boldsymbol{A}_1| = 0 \tag{20}$$

After reflection, similarly, Equation (21) is obtained

$$\boldsymbol{A}_2\boldsymbol{C}_2 = \boldsymbol{0} \tag{21}$$

Where, $\boldsymbol{C}_2 = (a_1, a_2, a_3, a_4, a_5, a_6, a_7, a_8)^{\mathrm{T}}$, $\boldsymbol{A}_2$ is an eight-order square matrix,

$$\boldsymbol{A}_2 = \begin{bmatrix} \boldsymbol{A}_{11} & \boldsymbol{A}_{12} \\ \boldsymbol{A}_{21} & \boldsymbol{A}_{22} \\ \boldsymbol{A}_{31} & \boldsymbol{A}_{32} \end{bmatrix} \tag{22}$$

Where

$$\boldsymbol{A}_{11} = \begin{bmatrix} \alpha_1^2 & 0 & \beta_1^2 & 0 \\ 0 & \alpha_1(-\frac{\xi^2\alpha_1^2}{4}+P) & 0 & \beta_1(-\frac{\xi^2\beta_1^2}{4}+P) \end{bmatrix}, \quad \boldsymbol{A}_{12} = \boldsymbol{A}_{31} = \boldsymbol{0}_{2\times 4},$$

$$\boldsymbol{A}_{21} = \begin{bmatrix} \cos\alpha_1(2-T) & \sin\alpha_1(2-T) & \cos\beta_1(2-T) & \sin\beta_1(2-T) \\ -\alpha_1\sin\alpha_1(2-T) & \alpha_1\cos\alpha_1(2-T) & -\beta_1\sin\beta_1(2-T) & \beta_1\cos\beta_1(2-T) \\ \alpha_1^2\cos\alpha_1(2-T) & \alpha_1^2\sin\alpha_1(2-T) & \beta_1^2\cos\beta_1(2-T) & \beta_1^2\sin\beta_1(2-T) \\ \alpha_1(\frac{\xi^2\alpha_1^2}{4}-P)\sin\alpha_1(2-T) & -\alpha_1(\frac{\xi^2\alpha_1^2}{4}-P)\cos\alpha_1(2-T) & \beta_1(\frac{\xi^2\beta_1^2}{4}-P)\sin\beta_1(2-T) & -\beta_1(\frac{\xi^2\beta_1^2}{4}-P)\cos\beta_1(2-T) \end{bmatrix}$$

$$\boldsymbol{A}_{32} = \begin{bmatrix} \cos\alpha_2 & \sin\alpha_2 & \cos\beta_2 & \sin\beta_2 \\ -\alpha_2\sin\alpha_2 & \alpha_2\cos\alpha_2 & -\beta_2\sin\beta_2 & \beta_2\cos\beta_2 \end{bmatrix},$$

$$\boldsymbol{A}_{22} = -\begin{bmatrix} \cos\alpha_2(2-T) & \sin\alpha_2(2-T) & \cos\beta_2(2-T) & \sin\beta_2(2-T) \\ -\alpha_2\sin\alpha_2(2-T) & \alpha_2\cos\alpha_2(2-T) & -\beta_2\sin\beta_2(2-T) & \beta_2\cos\beta_2(2-T) \\ \alpha_2^2\cos\alpha_2(2-T) & \alpha_2^2\sin\alpha_2(2-T) & \beta_2^2\cos\beta_2(2-T) & \beta_2^2\sin\beta_2(2-T) \\ \alpha_2(\frac{\xi^2\alpha_2^2}{4}-2P)\sin\alpha_2(2-T) & -\alpha_2(\frac{\xi^2\alpha_2^2}{4}-2P)\cos\alpha_2(2-T) & \beta_2(\frac{\xi^2\beta_2^2}{4}-2P)\sin\beta_2(2-T) & -\beta_2(\frac{\xi^2\beta_2^2}{4}-2P)\cos\beta_2(2-T) \end{bmatrix}$$

The bifurcation condition is

$$|\boldsymbol{A}_2| = 0 \tag{23}$$

5. Numerical results and analysis

The critical loads and buckling modes are predicted using MATLAB on the basis of the current mathematical model. Before and after reflection, the critical conditions of buckling are obtained from Equation (20) and Equation (23), respectively, which involve the following parameters: circumferential buckling mode n (including the axisymmetric mode 0 and non-axisymmetric modes 1, 2), the specific value of thickness to neutral surface radius of the cylindrical shell ξ, dimensionless buckling time T, dimensionless stiffness K, dimensionless damping $\overline{C}$, and the dimensionless parameters are adopted for generality and brevity of the analysis.

Before the stress wave reflection, Figure 2 presents the variation of the critical buckling load P with respect to the dimensionless radial damping $\overline{C}$ and stiffness K, respectively, based on Equation (20).

Figure 2 (a) shows the critical load P increasing with the radial damping $\overline{C}$ increasing, i.e., the radial damping constraints will enhance the anti-buckling capacity of cylindrical shells. The figure is demonstrated that critical loads P increasing with the radial stiffness K increasing, at a fixed value of the radial damping $\overline{C}$, i.e., the radial stiffness constraints will enhance the anti-buckling capacity of cylindrical shells. The larger the radial stiffness K is, the smaller the interval between the two adjacent curves is, which indicates that the radial stiffness has a much greater influence on the critical load at small radial stiffness. While the influence on critical load becomes gradually smaller as the radial stiffness increases.

Figure 2 (b) displays the larger the radial damping $\overline{C}$ is, the smaller the interval between the two adjacent curves is, which indicates that radial damping has a much greater influence on the critical load at small radial damping. Meanwhile, the influence on critical load becomes gradually smaller with the increase of radial damping.

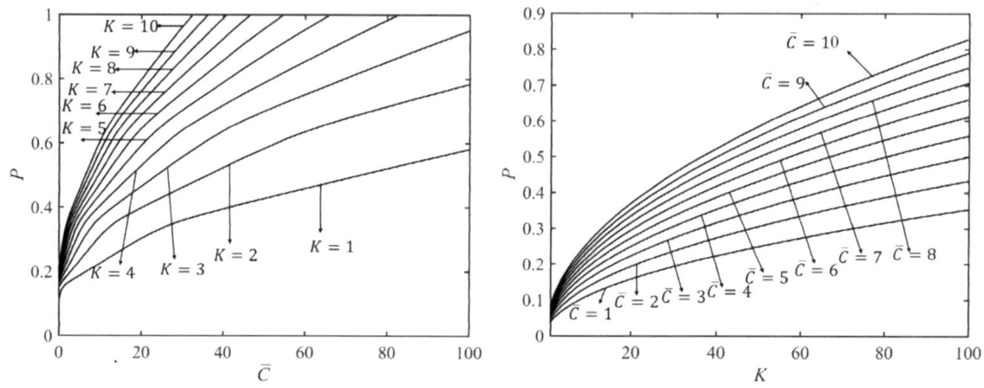

(a) (b)

Figure 2. Critical buckling load before reflection ($\xi = 0.05, T = 0.5, n = 0$)

Before the stress wave occurs reflection, Figure 3 presents the variation of critical loads P with respect to the buckling time T, when the shell constrained by several groups of radial stiffness and damping.

The curves in Figure 3 were obtained before the stress wave reflection. It reveals that the buckling critical loads P turn to decrease with the buckling time T growing, and then tends to be flat. This demonstrates that the critical load for the cylindrical shell approaching the static critical loads as the time of wave propagation grows very long. Figure 3 displays that the critical buckling load at $\overline{C}=1, K=10$ is higher than that at $\overline{C}=10, K=1$. The critical buckling load when $\overline{C}=1, K=100$ is higher than that at $\overline{C}=100, K=1$. And the critical buckling load when $\overline{C}=10, K=100$ is higher than that at $\overline{C}=100, K=10$. This manifests that the influence of radial stiffness on critical buckling load is greater than that of radial damping, in summary, the anti-buckling capability about the radial stiffness is more superior than radial damping.

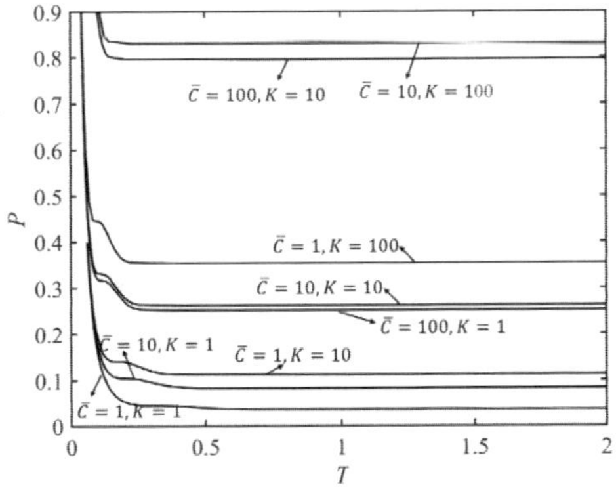

Figure 3. Critical buckling load vs. the buckling time before reflection ($\xi = 0.05, n = 0$)

Before the stress wave reflection, for $\xi = 0.05, T = 0.8, \bar{C} = 10$, Figure 4 (a) is the first third-order buckling mode of the cylindrical shell given by Equation (11), based on Equation (20). For $\xi = 0.05, T = 0.8$, $K = 10$, Figure 4 (b) is the first third-order buckling mode shapes about the cylindrical shell given by Equation (11), based on Equation (20). P corresponds to the critical loads of the cylindrical shell for different buckling modes. And the bottom of the cylindrical shell is the impact end.

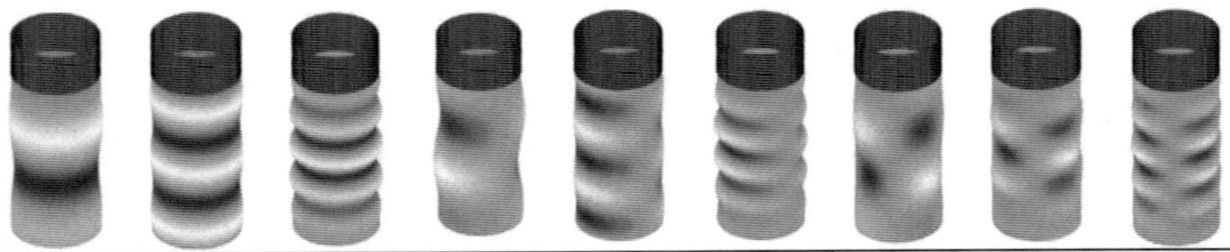

P	0.1928	0.5244	1.6936	0.1941	0.5257	1.6948	0.1981	0.5745	1.6986
K	1	10	100	1	10	100	1	10	100
n		0			1			2	

(a)

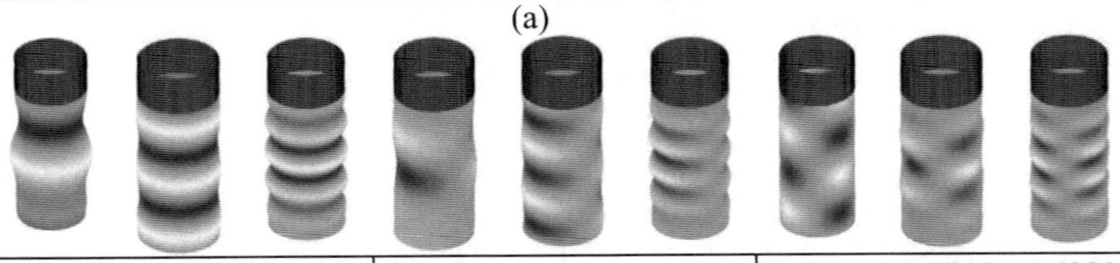

P	0.2694	0.5244	1.6230	0.2706	0.5257	1.6243	0.2787	0.5745	1.6280
$\bar{C}$	1	10	100	1	10	100	1	10	100
n		0			1			2	

(b)

Figure 4. Buckling modes of the shell with radial stiffness and damping constraints at $T = 0.8$

It can be noted from Figure 4, critical loads P increasing with the radial stiffness K and damping $\bar{C}$ increasing, and the corresponding axial buckling mode also increases, when value of n fixed. At fixed values about K and $\bar{C}$, the buckling critical loads P increases as the circumferential buckling mode n increases. This demonstrates that the axial and circumferential buckling modes increase as the critical load increases, before the stress wave reflection.

Figure 4 also shows the variation of the critical load caused by changes in the stiffness K is greater than in damping $\bar{C}$, at the same value of n. It is also demonstrated that the anti-buckling capability of the stiffness is more superior than the damping.

Local buckling occurs in the cylindrical shell withstand the axial impact, as the disturbance region of the stress wave grows, so does the buckling. Therefore, it can be seen from Figure 4, buckling only occurs in the stress wave disturbance region before stress wave occurs reflection.

After the stress wave occurs reflection, Figure 5 presents the variation of critical loads P with respect to the dimensionless radial damping $\bar{C}$ and stiffness K based on Equation (23).

Figure 5 (a) shows the critical loads P increasing with the radial damping $\bar{C}$ increasing; the critical loads P increasing with the radial stiffness K increasing. The larger the radial stiffness K is, the smaller the interval between the two adjacent curves is, which indicates that the radial stiffness has a much greater influence on the critical load at small radial stiffness. Meanwhile, the influence on critical load becomes gradually smaller as the radial stiffness increases.

Figure 5 (b) exhibits the larger the radial damping $\bar{C}$ is, the smaller the interval between the two adjacent curves is, which indicates that the radial damping has a much greater influence on the critical load at small radial damping. Meanwhile, the influence on critical load becomes gradually smaller with the increase of radial damping.

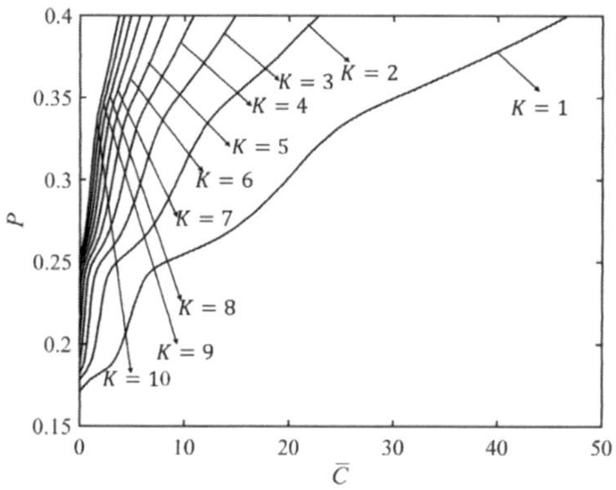

(a)

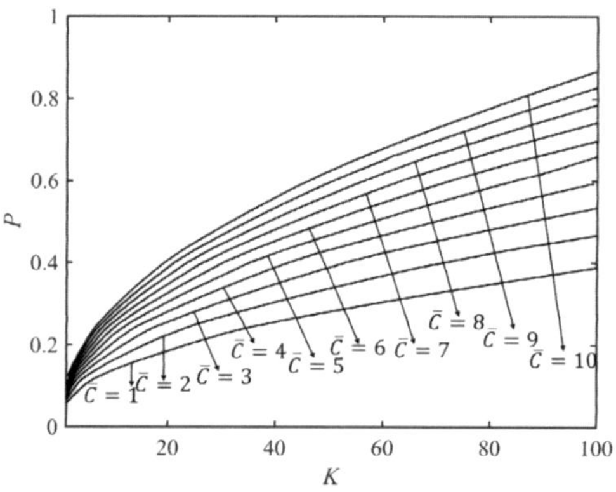

(b)

Figure 5. Critical buckling load after reflection ($\xi = 0.05, T = 1.5, n = 0$)

After the stress wave reflection, for $\xi = 0.05, T = 1.5, \bar{C} = 10$, Figure 6 (a) is the first third-order buckling mode of the cylindrical shell given by Equation (12), based on Equation (23). For $\xi = 0.05, T = 1.5$, $K = 10$, Figure 6 (b) is the first third-order buckling mode of the cylindrical shell given by Equation (12), based on Equation (23).

It can be noted from Figure 6, critical loads P increasing with the radial stiffness K and damping $\bar{C}$ increasing, and the corresponding axial buckling mode also increases, when value of n fixed. At fixed values of K and $\bar{C}$, the buckling critical loads P increases as the circumferential buckling mode n increases. This demonstrates that the axial and circumferential buckling modes increase as the critical buckling load increases, after the stress wave reflection.

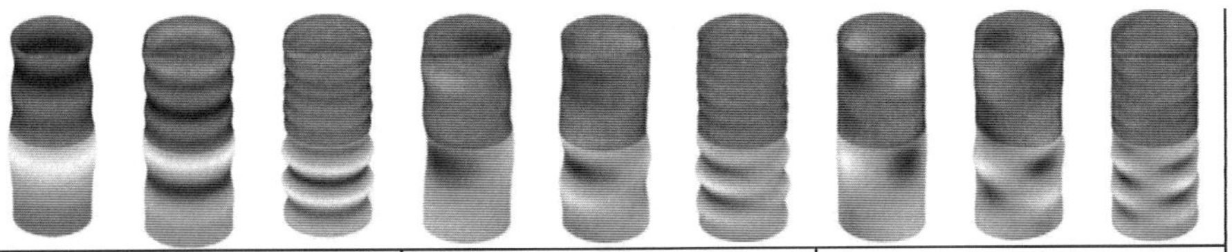

P	0.1780	0.5584	1.7242	0.1787	0.5683	1.7253	0.1810	0.5714	1.7286
K	1	10	100	1	10	100	1	10	100
n		0			1			2	

(a)

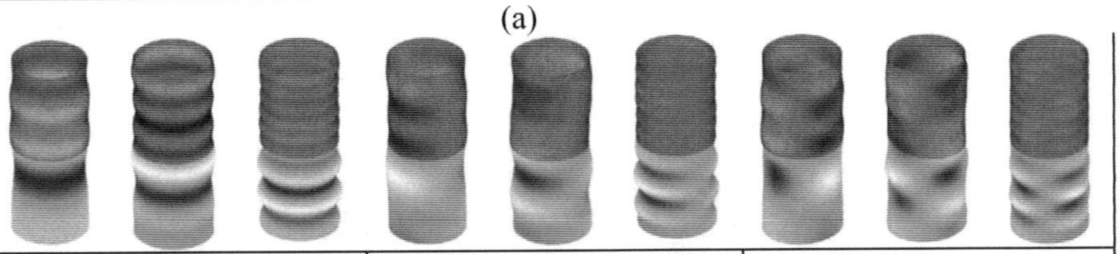

P	0.2493	0.5584	1.6637	0.2501	0.5683	1.6649	0.2524	0.5714	1.6684
$\bar{C}$	1	10	100	1	10	100	1	10	100
n		0			1			2	

(b)

Figure 6. Buckling modes of the shell with radial stiffness and damping constraints at $T=1.5$.

Figure 6 also exhibits that the variation for critical load caused by changes in the stiffness K is greater than that in damping $\bar{C}$ at the same value of n. It is also demonstrated that the anti-buckling capacity of the stiffness is better than the damping.

The whole cylindrical shell is divided into the stress wave disturbance region and the reflected wave disturbance region after stress wave reflection. Therefore, it could be concluded from Figure 6, the buckling covers the whole shell, and the axial buckling modes in the reflected wave disturbance region are denser than that in the stress wave disturbance region.

6. Conclusions

The dynamic buckling for cylindrical shells constrained by radial stiffness and damping withstand an axially impact is studied, the buckling critical loads and the mode shapes are derived, and the following conclusions are attained.

Both the radial stiffness and damping constraints will enhance the anti-buckling capacity of cylindrical shells. And the anti-buckling capacity of the former is better than the latter.

The radial stiffness and damping values will have a great influence on the critical load when they are small but do not have much influence on the critical load with their increasing values.

The axial and circumferential buckling modes increase as the critical load increases.

Before the stress wave occurs reflection, buckling only appears in the disturbance zone of the shell stress wave. After the stress wave occurs reflection, the shell buckling covers the entire region of a cylindrical shell, and axial buckling modes in the disturbance zone are denser than that in the undisturbed zone of the reflected stress wave.

References

[1] Srinivas T, Raju V V S and Narendar S 2013 Free vibration analysis of embedded circular cylindrical shells using wave propagation approach *International Journal of Research in Aeronautical and Mechanical Engineering* **1**

[2] Ji D 2008 Vibration analysis of cantilever cylindrical shell with magnetic constrained layer damping treatment *Chinese Journal of Solid Mechanics*

[3] Jin L and Sun S W 2014 Vibration analysis of short column shell with passive constrained layer

damping structure *Science and Technology Information*

[4] Gui Y F and Ma J M 2019 Buckling of step cylindrical shells under axial impact load *Journal of Vibration and Shock* **38** 200-5

[5] Gui Y F, Xin S J and Ma J M 2022 Dynamic characteristics calculation and analysis of axially impacted circular cylindrical shells with general boundary conditions *Journal of Vibration and Shock* **41** 143-51

Application of Multiscale Method in Nonlinear Vibration Problem of Multi-joint Manipulator

HaiYang Chang[1] and ShanYing Jiang[1*]

[1] School of Mechanical Engineering, Shanghai Institute of Technology, Shanghai 201418, China

18438621902@163.com jiangshanying@sit.edu.cn

Abstract: In this paper, the nonlinear two-joint manipulator system is taken as the research object. Firstly, the dynamic equation of the system is established by using the Lagrange method. In the case of ignoring some high-order nonlinear terms in the equations, the multi-scale method in the nonlinear vibration approximate analytical method is used to solve the nonlinear multi-joint manipulator system model, and the approximate analytical solution of the system is obtained. The theoretical analysis results are verified by numerical calculation. The results show that the analytical solution is in good agreement with the numerical solution, which verifies the effectiveness and accuracy of the multi-scale method.

1. Introduction

With the improvement of the level of industrial automation, the application of multi-joint manipulators in actual production is becoming more and more extensive, and the requirements for the structural design of manipulators are getting higher and higher. Not only a high load-to-weight ratio of manipulators is required, but also it is necessary to make the manipulator operate more flexibly, as well as have strong stability and safety.

However, the robot is a kind of typical nonlinear and strong coupling system, and it is difficult to obtain its accurate mathematical model. In addition, the working environment of the robot is becoming more and more complex, to better complete specific tasks, there are high requirements for trajectory tracking control of the manipulator. The existing robot trajectory tracking control methods generally include the adaptive control method, model-based control method, and PID control method [1-5].

In addition, due to the existence of joint flexibility, the manipulator will have residual vibration when the robot stops running, which will not only greatly reduce the working efficiency of the system, but also make the end of the manipulator unable to stay in the desired position quickly and accurately, so that the high-precision position control target is meaningless. Therefore, measures must be taken to restrain the vibration, and there are many research results in this area [6-9].

In this paper, the dynamic model of the two-joint nonlinear manipulator system is established, and the regular vibration law of each joint in the practical engineering application of the two-joint manipulator is studied. The approximate analytical solution of the system is obtained by using the multi-scale method [10], and the correctness of the theoretical analysis results is verified numerically.

2. The dynamic model of the two-joint manipulator

The research object of this paper is a two-joint rigid manipulator, and its simplified model is shown in Figure 1:

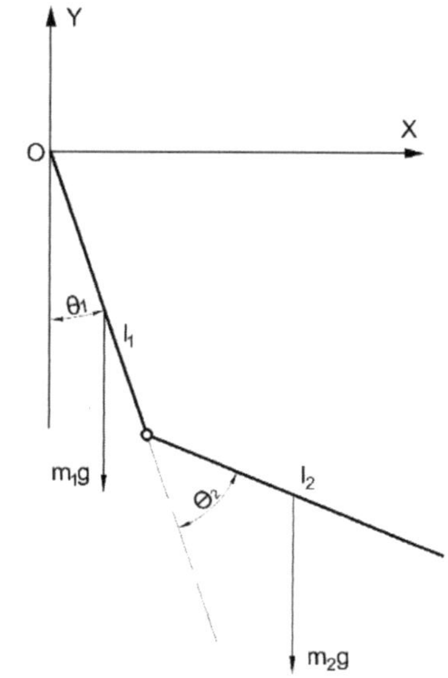

Figure 1. Structural model of the robotic arm system

The dynamic equations of the two-joint manipulator are established by the Lagrange method. Omitting the higher-order nonlinear terms in the equations, the dynamic model of the system can be obtained as:

$$\left.\begin{aligned}&\left[\left(\frac{1}{3}m_1+m_2\right)l_1^{\,2}+\frac{1}{3}m_2l_2^{\,2}+m_2l_1l_2\right]\ddot{\theta}_1+\left[\frac{1}{2}m_2l_1l_2+\frac{1}{3}m_2l_2^{\,2}\right]\ddot{\theta}_2\\&+\left(\frac{1}{2}m_1+m_2\right)gl_1\theta_1+\frac{1}{2}m_2gl_2(\theta_1+\theta_2)+c_1\dot{\theta}_1-\frac{1}{2}m_2l_1l_2\theta_2^{\,2}\ddot{\theta}_1\\&-\frac{1}{4}m_2l_1l_2\theta_2^{\,2}\ddot{\theta}_2-m_2l_1l_2\theta_2\dot{\theta}_1\dot{\theta}_2-\frac{1}{2}m_2l_1l_2\theta_2\dot{\theta}_2^{\,2}=F_1\cos\omega_1t\\&\frac{1}{3}m_2l_2^{\,2}\ddot{\theta}_2+\left(\frac{1}{3}m_2l_2^{\,2}+\frac{1}{2}m_2l_1l_2\right)\ddot{\theta}_1+\frac{1}{2}m_2gl_2(\theta_1+\theta_2)+c_2\dot{\theta}_2\\&-\frac{1}{4}m_2l_1l_2\theta_2^{\,2}\ddot{\theta}_1+\frac{1}{2}m_2l_1l_2\theta_2\dot{\theta}_1^{\,2}=F_2\cos\omega_2t\end{aligned}\right\}\tag{1}$$

3. Solution by the multi-scale method

The multi-scale method is a very effective approximate calculation method and is widely used in the solution of nonlinear differential equations. This paper will use this method to solve the approximate solution of the system model (1). The system of Equations (1) is first normalized:

$$\left.\begin{aligned}&\ddot{\theta}_1+\alpha_1\ddot{\theta}_2+\alpha_2\theta_1+\alpha_3(\theta_1+\theta_2)+k_1\dot{\theta}_1-2\alpha_4\theta_2^{\,2}\ddot{\theta}_1-\alpha_4\theta_2^{\,2}\ddot{\theta}_2-4\alpha_4\theta_2\dot{\theta}_1\dot{\theta}_2\\&-2\alpha_4\theta_2\dot{\theta}_2^{\,2}=f_1Cos\omega_1t\\&\ddot{\theta}_2+\beta_1\ddot{\theta}_1+\beta_2(\theta_1+\theta_2)+k_2\dot{\theta}_2-\beta_3\theta_2^{\,2}\ddot{\theta}_1+2\beta_3\theta_2\dot{\theta}_1^{\,2}=f_2Cos\omega_2t\end{aligned}\right\}\tag{2}$$

For weak nonlinear systems, to balance the effects of the external excitation, damping, and nonlinear terms, and make the effects of the external excitation, damping, and nonlinear terms appear in the same perturbation equation, a small dimensionless parameter ε is introduced, and the Equation group (2) can be transformed into:

$$\left.\begin{aligned}&\ddot{\theta}_1+\alpha_1\ddot{\theta}_2+\alpha_2\theta_1+\alpha_3(\theta_1+\theta_2)+\varepsilon\overline{k_1}\dot{\theta}_1-2\varepsilon\overline{\alpha_4}\theta_2{}^2\ddot{\theta}_1-\varepsilon\overline{\alpha_4}\theta_2{}^2\ddot{\theta}_2\\&-4\varepsilon\overline{\alpha_4}\theta_2\dot{\theta}_1\dot{\theta}_2-2\varepsilon\overline{\alpha_4}{}_4\theta_2\dot{\theta}_2{}^2=\varepsilon\overline{f_1}Cos\omega_1 t\\&\ddot{\theta}_2+\beta_1\ddot{\theta}_1+\beta_2(\theta_1+\theta_2)+\varepsilon\overline{k_2}\dot{\theta}_2-\varepsilon\overline{\beta_3}\theta_2{}^2\ddot{\theta}_1+2\varepsilon\overline{\beta_3}\theta_2\dot{\theta}_1{}^2=\varepsilon\overline{f_2}Cos\omega_2 t\end{aligned}\right\}\tag{3}$$

In the equation, $\overline{k_1}=\frac{k_1}{\varepsilon}$, $\overline{\alpha_4}=\frac{\alpha_4}{\varepsilon}$, $\overline{f_1}=\frac{f_1}{\varepsilon}$, $\overline{k_2}=\frac{k_2}{\varepsilon}$, $\overline{\beta_3}=\frac{\beta_3}{\varepsilon}$, $\overline{f_2}=\frac{f_2}{\varepsilon}$

Now we solve a solution of the form shown below:

$$\left.\begin{aligned}\theta_1&=\theta_{10}(T_0,\ T_1)+\varepsilon\theta_{11}(T_0,\ T_1)+\cdots\\\theta_2&=\theta_{20}(T_0,\ T_1)+\varepsilon\theta_{21}(T_0,\ T_1)+\cdots\end{aligned}\right\}\tag{4}$$

Among them $T_1=\varepsilon T_0$, at the same time: $\frac{d}{dt}=D_0+\varepsilon D_1$, $\frac{d^2}{dt^2}=D_0^2+2\varepsilon D_0D_1+\varepsilon^2D_1^2$.

By substituting the above equation and Equation (4) into the system Equation (3) and setting the coefficient of the same power of ε to be zero, linear partial differential equations of each order approximation are obtained as follows:

$$\varepsilon^0:\quad\left.\begin{aligned}&D_0{}^2\theta_{10}+\alpha_1D_0{}^2\theta_{20}+\alpha_2\theta_{10}+\alpha_3(\theta_{10}+\theta_{20})=0\\&D_0{}^2\theta_{20}+\beta_1D_0{}^2\theta_{10}+\beta_2(\theta_{10}+\theta_{20})=0\end{aligned}\right\}\tag{5}$$

$$\varepsilon^1:\quad\left.\begin{aligned}&D_0{}^2\theta_{11}+\alpha_1D_0{}^2\theta_{21}+\alpha_2\theta_{11}+\alpha_3(\theta_{11}+\theta_{21})=-2D_0D_1\theta_{10}-2\alpha_1D_0D_1\theta_{20}-\overline{k_1}D_0\theta_{10}+\\&2\overline{\alpha_4}\theta_{20}{}^2D_0{}^2\theta_{10}+\overline{\alpha_4}\theta_{20}{}^2D_0{}^2\theta_{20}+4\overline{\alpha_4}\theta_{20}(D_0\theta_{10}D_0\theta_{20})+2\overline{\alpha_4}\theta_{20}(D_0\theta_{20})^2+\overline{f_1}Cos\omega_1t\\&D_0{}^2\theta_{21}+\beta_1D_0{}^2\theta_{11}+\beta_2(\theta_{11}+\theta_{21})=-2D_0D_1\theta_{20}-2\beta_1D_0D_1\theta_{10}-\overline{k_2}D_0\theta_{20}+\\&\overline{\beta_3}\theta_{20}{}^2D_0{}^2\theta_{10}-2\overline{\beta_3}\theta_{20}(D_0\theta_{10})^2+\overline{f_2}Cos\omega_2t\end{aligned}\right\}\tag{6}$$

The solution to Equation (5) is as follows:

$$\left.\begin{aligned}\theta_{10}&=A(T_1)\exp(i\omega_{11}T_0)+B(T_1)\exp(i\omega_{12}T_0)+cc\\\theta_{20}&=\mu_1A(T_1)\exp(i\omega_{11}T_0)+\mu_2B(T_1)\exp(i\omega_{12}T_0)+cc\end{aligned}\right\}\tag{7}$$

Where cc denotes the conjugate complex of the terms on the left.

From the linear vibration theory, the following formulas can be obtained:

$$\omega_{1i}{}^2=\frac{(\alpha_2+\alpha_3+\beta_2-\alpha_1\beta_2-\alpha_3\beta_1)\mp\sqrt{(\alpha_2+\alpha_3+\beta_2-\alpha_1\beta_2-\alpha_3\beta_1)^2-4(1-\alpha_1\beta_1)\alpha_2\beta_2}}{2(1-\alpha_1\beta_1)}$$

$$\mu_i=\frac{\beta_2-\beta_1\omega_{1i}{}^2}{\omega_{1i}{}^2-\beta_2},\quad i=1,2\tag{8}$$

By substituting Equation (7) into Equation (6), we can obtain:

$$\begin{aligned}&D_0{}^2\theta_{11}+\alpha_1D_0{}^2\theta_{21}+\alpha_2\theta_{11}+\alpha_3(\theta_{11}+\theta_{21})=\\&R_{11}\exp(i\omega_{11}T_0)+R_{21}\exp(i\omega_{12}T_0)+\frac{1}{2}\overline{f_1}\exp(i\omega_1T_0)+NST\end{aligned}\tag{9}$$

$$\begin{aligned}&D_0{}^2\theta_{21}+\beta_1D_0{}^2\theta_{11}+\beta_2(\theta_{11}+\theta_{21})=\\&R_{12}\exp(i\omega_{11}T_0)+R_{22}\exp(i\omega_{12}T_0)+\frac{1}{2}\overline{f_2}\exp(i\omega_2T_0)+NST\end{aligned}\tag{10}$$

Where *NST* represents the non-duration term and the term that may cause internal resonance, and the other coefficients are:

$$R_{11}=-2i\omega_{11}(1+\alpha_1\mu_1)A'-\overline{k_1}i\omega_{11}A-2\overline{\alpha_4}\mu_2{}^2\omega_{11}{}^2(2+\mu_1)AB\overline{B}-2\overline{\alpha_4}\mu_1{}^2\omega_{11}{}^2\left(1+\frac{1}{2}\mu_1\right)A^2\overline{A}$$

$$R_{21}=-2i\omega_{12}(1+\alpha_1\mu_2)B'-\overline{k_1}i\omega_{12}B-2\overline{\alpha_4}\mu_1{}^2\omega_{12}{}^2(2+\mu_2)A\overline{A}B-2\overline{\alpha_4}\mu_2{}^2\omega_{12}{}^2\left(1+\frac{1}{2}\mu_2\right)B^2\overline{B}$$

$$R_{12}=-2i\omega_{11}(\mu_1+\beta_1)A'-\overline{k_2}i\omega_{11}\mu_1A-2\overline{\beta_3}\left(\mu_2{}^2\omega_{11}{}^2+2\mu_1\mu_2\omega_{12}{}^2+2\mu_1\omega_{12}{}^2\right)AB\overline{B}-$$
$$2\overline{\beta_3}\left(\frac{3}{2}\mu_1{}^2\omega_{11}{}^2+\mu_1\omega_{11}{}^2\right)A^2\overline{A}$$

$$R_{22}=2i\omega_{12}(\mu_2+\beta_1)B'-\overline{k_2}i\omega_{12}\mu_2B-2\overline{\beta_3}\left(\mu_1{}^2\omega_{12}{}^2+2\mu_1\mu_2\omega_{11}{}^2+2\mu_2\omega_{11}{}^2\right)A\overline{A}B-$$
$$2\overline{\beta_3}\left(\frac{3}{2}\mu_2{}^2\omega_{12}{}^2+\mu_2\omega_{12}{}^2\right)B^2\overline{B}$$

Supposing $\omega_1=\omega_{11}+\varepsilon\sigma_1$, $\omega_2=\omega_{12}+\varepsilon\sigma_2$, to obtain the solvable conditions from Equations (9) and (10), it is assumed that Equations (9) and (10) have the following special solutions:

$$\left.\begin{aligned}\theta_{11}&=A_1\exp(i\omega_{11}T_0)+B_1\exp(i\omega_{12}T_0)\\ \theta_{21}&=A_2\exp(i\omega_{11}T_0)+B_2\exp(i\omega_{12}T_0)\end{aligned}\right\}\tag{11}$$

Substituting it into Equations (9) and (10), and making the coefficients of $\exp(i\omega_{11}T_0)$ and $\exp(i\omega_{12}T_0)$ equal, we can get:

$$\left.\begin{aligned}&(\alpha_2+\alpha_3-\omega_{11}{}^2)A_1+(\alpha_3-\alpha_1\omega_{11}{}^2)A_2=R_{11}+\frac{1}{2}\overline{f_1}\exp(i\sigma_1T)\\ &(\beta_2-\beta_1\omega_{11}{}^2)A_1+(\beta_2-\omega_{11}{}^2)A_2=R_{12}\end{aligned}\right\}\tag{12}$$

$$\left.\begin{aligned}&(\alpha_2+\alpha_3-\omega_{12}{}^2)B_1+(\alpha_3-\alpha_1\omega_{12}{}^2)B_2=R_{21}\\ &(\beta_2-\beta_1\omega_{12}{}^2)B_1+(\beta_2-\omega_{12}{}^2)B_2=R_{22}+\frac{1}{2}\overline{f_2}\exp(i\sigma_2T_1)\end{aligned}\right\}\tag{13}$$

According to the knowledge of linear algebra theory, if A1, A2, B1, and B2 in linear nonhomogeneous equations have nonzero solutions, then the rank of the coefficient matrix and augmented matrix must be the same. From the calculation results of ω_{11} and ω_{12} in Equation (8), it can be known that the coefficient determinants of Equations (12) and (13) are zero, so the necessary and sufficient conditions for A1, A2, B1, B2 to have solutions is that the following determinants are zero:

$$\begin{vmatrix}\alpha_3-\alpha_1\omega_{11}{}^2 & R_{11}+\frac{1}{2}\overline{f_1}\exp(i\sigma_1T_1)\\ \beta_2-\omega_{11}{}^2 & R_{12}\end{vmatrix}=0\tag{14}$$

$$\begin{vmatrix}\alpha_3-\alpha_1\omega_{12}{}^2 & R_{21}\\ \beta_2-\omega_{12}{}^2 & R_{22}+\frac{1}{2}\overline{f_2}\exp(i\sigma_2T_1)\end{vmatrix}=0\tag{15}$$

Therefore, the solvable conditions are as follows:

$$\left.\begin{aligned}&(\beta_2 - \omega_{11}^2)\left[R_{11} + \frac{1}{2}\overline{f_1}\exp(i\sigma_1 T_1)\right] + (\alpha_1\omega_{11}^2 - \alpha_3)R_{12} = 0\\&(\beta_2 - \omega_{12}^2)R_{21} + (\alpha_1\omega_{12}^2 - \alpha_3)\left[R_{22} + \frac{1}{2}\overline{f_2}\exp(i\sigma_2 T_1)\right] = 0\end{aligned}\right\} \tag{16}$$

Equation (16) can be simplified as:

$$\left.\begin{aligned}&iA' + ip_{11}A + p_{12}A^2\overline{A} + p_{13}AB\overline{B} + p_{14}\overline{f_1}\exp(i\sigma_1 T_1) = 0\\&iB' + ip_{21}B + p_{22}B^2\overline{B} + p_{23}A\overline{A}B + p_{24}\overline{f_2}\exp(i\sigma_2 T_1) = 0\end{aligned}\right\} \tag{17}$$

To make it easier to solve, let:

$$\left.\begin{aligned}&A = \frac{1}{2}a(T_1)\exp[i\phi_1(T_1)]\\&B = \frac{1}{2}b(T_1)\exp[i\phi_2(T_1)]\end{aligned}\right\} \tag{18}$$

Substituting the above equation into Equation (17), and separating the real and imaginary parts, Equation (17) can be transformed into the following system of autonomous differential equations:

$$\left.\begin{aligned}&a' = -p_{11}a + 2p_{14}\overline{f_1}Sin\varphi_1\\&a\varphi_1' = \frac{1}{4}p_{12}a^3 + \frac{1}{4}p_{13}ab^2 + 2p_{14}\overline{f_1}Cos\varphi_1 - a\sigma_1\\&b' = -p_{21}b + 2p_{24}\overline{f_2}Sin\varphi_2\\&b\varphi_2' = \frac{1}{4}p_{22}b^3 + \frac{1}{4}p_{23}a^2b + 2p_{24}\overline{f_2}Cos\varphi_2 - b\sigma_2\end{aligned}\right\} \tag{19}$$

Among them, $\varphi_1 = \emptyset_1 - \sigma_1 T_1$, $\varphi_2 = \emptyset_2 - \sigma_2 T_1$. According to whether $\overline{f_i}(i = 1,2)$ is zero or not, the system can be divided into three cases. This article only discusses the casc where $\overline{f_2}$is zero.

Settig $a' = b' = \varphi_1' = \varphi_2' = 0$, we obtain the steady-state solution of the system (19). From the system of Equations (19) we know that $b = 0$. So, the steady-state solution of system (19) depends on the following two-dimensional system:

$$\left.\begin{aligned}&a' = -p_{11}a + 2p_{14}\overline{f_1}Sin\varphi_1\\&\varphi_1' = \frac{1}{4}p_{12}a^2 + 2p_{14}\overline{f_1}a^{-1}Cos\varphi_1 - \sigma_1\end{aligned}\right\} \tag{20}$$

Let $a' = \varphi_1' = 0$, the steady-state solution of Equation (20) can be obtained as:

$$b = 0,\ 64p_{14}^2\overline{f_1}^2 = 16p_{11}^2a^2 + (p_{12}a^3 - 4a\sigma_1)^2 \tag{21}$$

Noting that $\omega_1 = \omega_{11} + \varepsilon\sigma_1$, the first approximate analytical solution of the system is:

$$\left.\begin{aligned}&\theta_{10} = aCos(\varphi_1 + \omega_1 t)\\&\theta_{20} = \mu_1 aCos(\varphi_1 + \omega_1 t)\end{aligned}\right\} \tag{22}$$

4. Numerical calculation and analysis

This paper mainly studies the law of vibration of each joint of the multi-joint manipulator in practical engineering applications. Now take the following data as an example for analysis and numerical calculation: $l_1 = 1m$, $l_2 = 2.5m$, $m_1 = 78kg$, $m_2 = 36kg$, $\alpha_1 = 0.53$, $\alpha_2 = 3.24$, $\alpha_3 = 1.94$,

$\alpha_4 = 0.10$, $\beta_1 = 1.6$, $\beta_2 = 5.88$, $\beta_3 = 0.30$, $\varepsilon = 0.1$, $\sigma = 0.1$. At high-frequency resonance, $\omega_{11} = \omega_{12} = 5.18$, then $\omega_1 = \omega_2 = 5.19$. When $f_1 = 0.5$, $f_2 = 0$, take $k_1 = 0.7$, and $k_2 = 0.1$. Figure 2 shows the motion law of the angular displacement θ of the connecting rod of the two-joint manipulator system as a function of time. The solid line in the figure represents the analytical solution, and the dashed line represents the numerical solution. Figures 3 and 4 show the phase plane curves of the angular displacements θ_1 and θ_2 in Figure 2, respectively. It can be seen from the following results that the approximate analytical solution obtained by the multi-scale method is in good agreement with the numerical solution of the system.

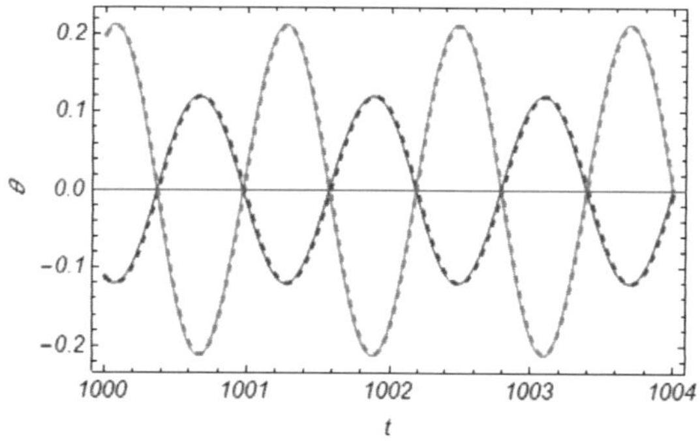

Figure 2. Comparison of analytical and numerical solutions of θ_1 and θ_2

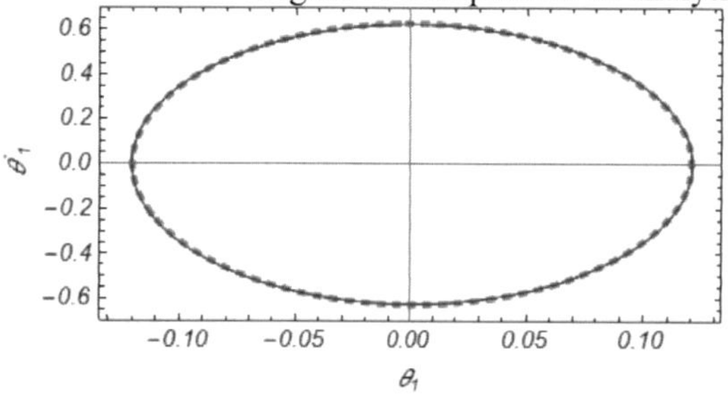

Figure 3. Comparison of phase plane curves of θ_1 and $\dot{\theta}_1$

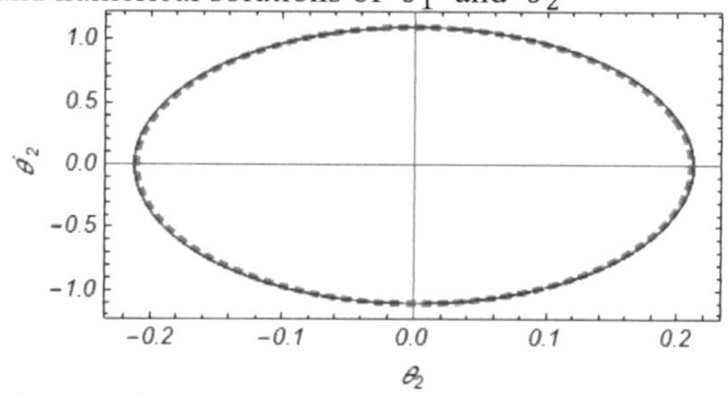

Figure 4. Comparison of phase plane curves of θ_2 and $\dot{\theta}_2$

5. Conclusion

In this paper, the author establishes the dynamic model of the two-joint manipulator system, obtains the approximate analytical solution of the system by using the multi-scale method, and verifies it numerically. The results show that the analytical solution is in good agreement with the numerical solution, which proves the correctness and effectiveness of the multi-scale method in solving the approximate solution of the multi-joint manipulator system.

References

[1] Qian Q, Zhang A H and Sun Y X 2019 *J. Ordn. Eng.* **40**(8):1732-1739
[2] Li Z N, Zhang J, Yin Y F, Wu K Y and Sun X G 2021 *Mech. Sci. Technol.* **40**(02):270-275
[3] Chen L Q and Zeng Z Z 2022 *Space. Contr. Technol. Appl.* **48**(04):70-77
[4] Ding R L, Hou X and Sun W J 2022 *Mech. Elec. Eng.* **39**(03):291-299
[5] Zhang J, Li Z N, Yin Y F, Qi C and Wu K Y 2021 *Mech. Sci. Technol.* **40**(05):710-715
[6] Cui S P, Sun Y J, Liu Y W and Liu H 2021 *Tr. Nanjing. U. Aero. Astro.* **38**(03):361-372
[7] Liu K Y and Yang M 2022 *J. Elec. Mach. Ctrl.* **26**(07):20-28
[8] Jia P X, Zhou Y and Li X W 2018 *Vib. Shock.* **37**(13):177-181
[9] Rong X P and Zhang D L 2015 *Mech. Elec. Eng.* **32**(01):14-20
[10] Liu Y Z and Chen L Q 2001 *nonlinear vibration* (Beijing: Higher Education Press) p 83

Research on Drag Reduction Flow Control Technology for High-Speed EMUs

Na Gao[1,2*], Jian Du[3]

[1]AVIC Aerodynamics Research Institute, Harbin, 150001, China,

[2]Low Speed High Reynolds Number Aeronautical Key Laboratory, Harbin, 150001, China,

[3]China Railway Rolling Stock Corporation Qingdao Sifang Co., Ltd, Qingdao, 266000, China,

[*]Corresponding author's e-mail:gn0427@163.com

Abstract. In order to explore a new method of aerodynamic drag reduction for high-speed trains, the numerical calculation method based on SST k-ω was used to explore the influence of jets at different positions of the train body, joints, tail and different speeds on the turbulent flow field and aerodynamic drag around the train. The research results show that: the whole train with an all-inclusive windshield reduces the drag by about 8.17%; the bionic groove can reduce the drag by about 2.53%; there is a separate area at the rear of the train, and a partial air-inhalation drag reduction scheme is proposed. By calculation, the maximum drag reduction rate is 3.84%. The drag reduction scheme of the vortex generator is arranged locally. Through the calculation of 30 schemes, the maximum drag reduction rate is 3.25%.

1. Introduction

High-speed railway has become the focus of the development of countries around the world for their advantages of high efficiency, convenience, comfort, safety, and a green environment. However, with the continuous improvement of train speed, passenger capacity, and transport distance, the energy consumption of trains has increased greatly, which poses new challenges to energy conservation and environmental protection. In order to ensure the sustainable development of high-speed trains and build a resource-saving and environment-friendly society, it is urgent to study the energy-saving and consumption reduction of trains.

Energy consumption during train operation is mainly used to overcome running resistance. In high-speed operation, the air resistance of trains increases sharply with the square of the speed. Research on the air resistance of trains is the main consideration for the drag reduction and energy saving of high-speed trains. In the research process of aerodynamic drag reduction of railway trains, with the development of computer technology, a large number of scholars are devoted to the research of high-speed train head design [1-6] and local structure optimization so as to achieve the purpose of drag and noise reduction. A series of substantial research results have been achieved, which greatly reduces the train running resistance. However, the traditional method to improve the aerodynamic performance of trains by changing the shape has become more and more perfect and mature and is limited by the continuous improvement of manufacturing technology and design requirements, which gradually shows its limitation. It is difficult to make a breakthrough. Therefore, scholars began to explore new drag-reduction methods.

The drag reduction technology of turbulent flow field active control is a cutting-edge research field in the world. It changes the local and even global flow field structure around the object by applying appropriate energy excitation on the premise of not changing the shape of the object so as to optimize the aerodynamic performance. This method has been widely used in aerospace and other fields, but it is only in the initial stage of exploration in ground vehicles. Large eddy simulation was used to study the wake characteristics of the 25°Ahmed model, and an unsteady jet active control method based on half sine wave signal was proposed, which has a more effective drag reduction effect compared with the traditional synthetic jet method. SST k-ω turbulence model was used to study steady-state jet drag reduction of the standard 25°Ahmed model. Jet holes were set at the tail of the model to analyze the influence rules of jet hole aperture, shape, jet velocity, jet angle, and other parameters on the aerodynamic drag of the train. The optimal jet scheme achieved a good drag reduction of 13.23%. Based on the Boltzmann method, jet flow control was carried out on the wake structure of the blunt body, and jet holes were arranged on the boundary of the vertical plane of the tail end of the blunt body. The results showed that when the jet flow and the incoming flow were 45° and pointed to the ground, the drag reduction effect was significant. When the jet flow velocity was 0.5 times the incoming flow velocity, the drag reduction rate was as high as 20%.

However, there is much research on the aerodynamic drag reduction of high-speed trains based on the flow control technology of turbulence drag reduction. In this paper, a numerical calculation method is adopted to simulate the flow field structure and aerodynamic drag characteristics around high-speed EMUS and to explore the installation of drag reduction measures in the special flow parts of the body and tail. The influences of jet position, jet velocity, spoiler position and size, bionic groove shape and size on the flow field structure, and aerodynamic drag of high-speed trains were studied, and the optimal drag reduction pattern was obtained. The research results of this paper provide a theoretical basis for the research of new drag reduction technology of high-speed trains and contribute to the sustainable development of high-speed railway energy saving and consumption reduction.

2. Mathematical model

2.1. Control equation

The train follows the laws of conservation of mass, momentum, and energy when moving. The governing equation in its general form can be written as:

$$\frac{\partial(\rho\phi)}{\partial t}+div(\rho\bar{u}\phi)=div(\Gamma grad\phi)+S \tag{1}$$

Where Φ is a general variable, Γ is a generalized diffusion coefficient, and S is a generalized source term. In Formula 1, the terms are transient term, convection term, diffusion term, and source term.

In order to close the above equations, it is necessary to introduce the gas equation of state:

$$P=\rho RT \tag{2}$$

Where R is the gas constant; T is the temperature of the gas.

2.2. Initial conditions and boundary conditions

Initial conditions such as density, initial velocity, temperature, and pressure of the flow field are given according to the calculation requirements. The boundary conditions are set as follows:

(1) Pressure far-field boundary conditions are adopted at the upper and side boundaries of the air region. The pressure far-field boundary condition is a reflection-free boundary based on the Riemann invariant. The pressure wave generated in the calculated region is not a reflected wave but a transmitted wave.

(2) The wall of the train adopts no-slip boundary conditions; that is, the normal velocity of the fluid is zero.

(3) Standard wall function is used to simulate the influence of tunnel wall roughness.

2.3. Solution

In this paper, the SST k-ω turbulence model is used, which has good performance for rotational flows, boundary layer flows with a strong adverse pressure gradient, flow separation, and secondary flows. The pressure-velocity coupling was solved using the SIMPLEC algorithm. The pressure term of the governing equation was discretized using the second-order upwind scheme. The momentum, turbulent kinetic energy, and dissipation rate were all in the QUICK scheme. Residual items are set to 10-6 to ensure the accuracy of numerical simulation.

2.4. Numerical method validation

In order to verify the correctness of the numerical calculation, the intersection pressure wave shape measured by the real vehicle test at 200 km/h is compared with the numerical calculation results. Figure 1 shows the head wave curve of a real train test and the head wave curve of numerical calculation at the measuring point in the middle of the train. It can be seen that the numerical calculation is basically consistent with the law obtained in the real vehicle test. The amplitude of the head wave and the tail wave are △P = 606Pa and △P = 372Pa, respectively. The measured data of the meeting pressure wave test in the test are as follows: the head wave amplitude is △P = 640Pa, and the tail wave amplitude is △P = 391Pa. The calculation results show that the relative error of the two is about 5.31% in the "head-to-head" intersection and about 5% in the "head-to-tail" intersection. The error range meets the engineering requirements, making the calculation results reliable.

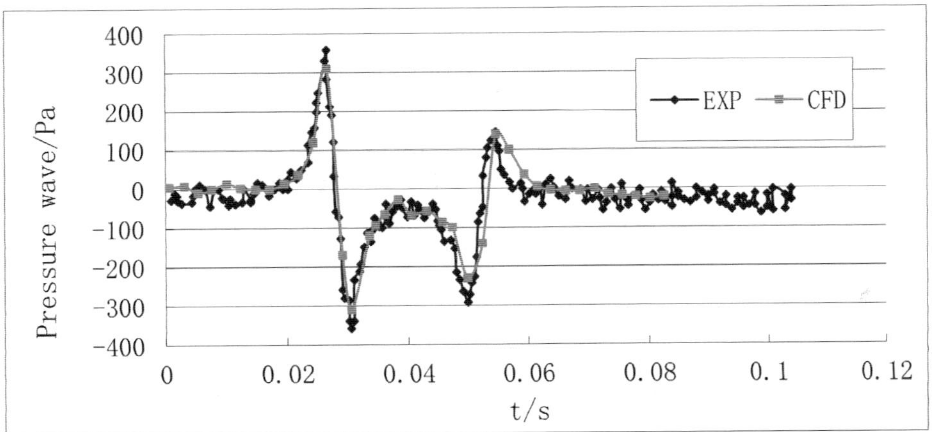

FIG. 1 Comparison of CFD and EXP results of EMU isometric rendezvous

3. Results and analysis

3.1. Analysis of original train calculation results

The above verified numerical simulation method is used to simulate the outflow field of the train under the condition of 500 km/h. By analyzing the resistance value of each train component, the optimization area is found, and the optimization scheme is proposed.

After the train speed increases, the airflow speed on the front and rear surfaces of the train intensifies, forming a strong impact in the front and a serious wake in the rear. The airflow condition in the rear of the train is shown in Figure 2. In the rear area of the train, the velocity of the airflow increases obviously at the inflection point of the curved surface, and the static pressure of the train body decreases. As the airflow approaches the nose end of the train tail, the speed decreases, and the static pressure on the train body surface increases. However, due to the energy dissipation caused by the flow separation in the flow field and the frictional resistance of the train body wall, the maximum pressure on the train tail is less than that on the head.

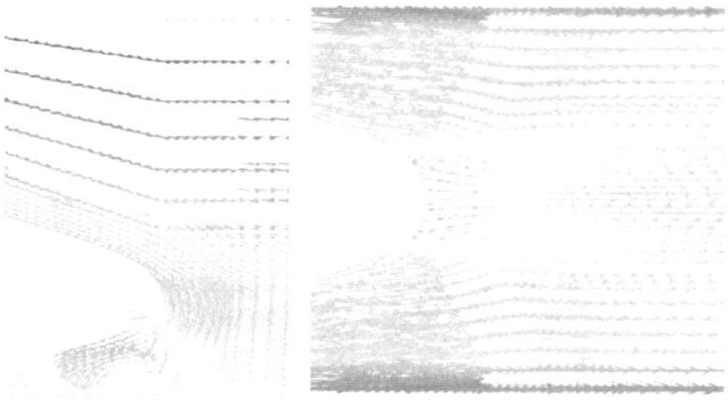

Fig2 Vector diagram of train exhaust flow velocity

As seen in Figure 2, due to the fast speed of the train, the airflow forms a serious eddy in the wake of the train, and the influence range is far away, lasting dozens of meters from the rear of the train. The existence of an eddy current not only affects the stability of the whole flow field but also increases the negative pressure of the rear of the train and increases the resistance value of the train. At the same time, the disordered disturbance and eddy current of the airflow near the wall of the train also interfere with the stability of the whole flow field and affect the pressure and resistance value of each part and part of the train, thus increasing the resistance value of the train.

From the above analysis, it can be seen that many areas on the train surface can be optimized, and the reasons for the higher aerodynamic resistance of the train-building train are mainly as follows.

(1) There is a large gap at the train body connection, where the forward direction of the airflow changes, forming eddy currents. The existence of an eddy current increases the drag of both front and rear trains. The optimization of the joint should try to prevent the airflow from being interrupted here and maintain the airtight surface of the train.

(2) There is a large separation zone in the tail flow, which should be blown away.

(3) There is a large separation area in the tail flow, so the eddy current generator and other diversion devices can be arranged for drag reduction research.

(4) Since the body area is large and the length is long, the research on reducing friction resistance should be focused on this part.

3.2. Bionic non-smooth groove drag reduction

The triangular groove with a good drag reduction effect was selected for this calculation, and its shape is shown in Figure 3; h is the height, and s is the distance between the humps.

For FIG. 3 (a), when h = 75μm, s = 130, 140, 150μm; When h = 200μm, s = 400μm. For Figure 3 (b), the triangle height is 75μm, and five states are calculated.

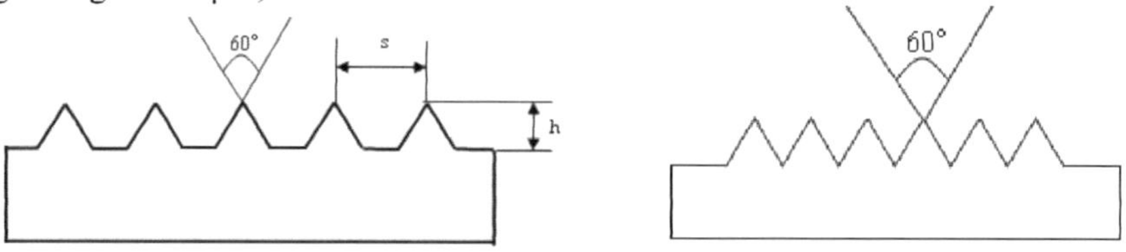

(a) Discontinuous arrangement of triangles (b) Continuously arranged triangles

FIG. 3 Schematic diagram of groove section geometry

Considering the running speed of the train is 500km/h (138.89m/s), only the drag reduction rate at the flow speed of 140m/s is calculated for the grooves of 75*130, 75*140, and 75*75 sizes. The results are shown in Table 1.

Table 1 Drag reduction effect under different sizes at a speed of 140 m/s

Size (μm)	Smooth surface resistance coefficient	Trench surface resistance coefficient	Drag reduction rate
75*140	0.070387422	0.064228736	8.75%
75*130	0.068979065	0.063223195	8.34%
75*75	0.081883514	0.064509875	21.22%

The corresponding body friction resistance accounts for 28.94% of the total resistance. It can be seen that the maximum drag reduction rate of the bionic groove applied to the train body of the three sections is 2.53%.

3.3. Full package outside the windshield drag reduction

Through the analysis of the original car model, it is found that because the inner windscreen only connects the two car bodies, there is a large gap between the car bodies. The airflow along the direction of the car body will change here. Some of the airflows continue to move forward, and some of the airflows enter the space between the car bodies and impact the rear end wall, forming a red high-pressure zone. Blocked by the rear end wall, the airflow flows backward, creating a vortex area. The flow of gas in the vortex zone affects the pressure distribution of the end wall and the inner windshield surface of both vehicles and increases the resistance value of the vehicle.

Therefore, in order to solve this adverse effect, the upper and lower gaps of the windshield are filled with soft materials to make the whole outer windshield. Except for the gap left below the car body, the outer windshield completely covers the upper and side of the inner windshield so that the front and rear car bodies achieve a smooth transition without the gap. Call it the "all-inclusive windshield model." The windshield structure is shown in Figure 4.

Original train model All-inclusive outside windshield model
FIG. 4 Model comparison after modification of windshield

With the addition of the all-inclusive windshield, there is no gap connection between the car bodies, and the airflow can be smoothly transferred from the front car to the rear car without causing eddy currents between the car bodies nor causing direct impact on the windward end wall of the rear car.

After the external windshield is added, the airflow at the connection of the two cars has no gap flow, which reduces the pressure and resistance value at the windshield and the rear car's resistance value, thus reducing the resistance value of the vehicle.

When the windshield is not fully covered, the vehicle resistance value is 0.36389, and when the windshield is added, the vehicle resistance value is 0.33413, and the vehicle drag reduction rate is 8.17%.

3.4. Research on drag reduction of nose and tail by blowing and inhaling

For the first time, the drag reduction calculation of the tail position under the nose tip was carried out, and the blowing speed was 150 m/s, which had no drag reduction effect, so it was not advisable. For the second time, the suction drag reduction calculation of the lower tip of the nose at the rear position was carried out, and the suction speed was 150 m/s, which had no drag reduction effect, so it was not advisable. For the third time, the drag reduction calculation of the head position was carried out. The blowing speed was 150m/s, and the drag reduction effect was achieved in both cases. The drag reduction was about 50%, which can be further analyzed. For the fourth time, the suction drag reduction calculation of the front position was carried out. The suction speed was 150m/s, but there was no drag reduction effect, so it was not advisable. For the fifth time, drag reduction calculation was carried out at the tail position above the nose tip at the blowing speed of 10~150 m/s (interval of 10 m/s) and 200 m/s, all of which had a drag reduction effect and could be further analyzed.

As can be seen from the above calculation results:

(1) There is no drag reduction effect on the nose and tail inhalation.

(2) The head blowing has a drag reduction effect, and the greater the blowing speed, the better the drag reduction effect, whether it is available to further analysis to determine.

(3) The upper part of the nose tip at the rear position has a drag reduction effect when the blowing speed is less than 120m/s, the maximum drag reduction speed is 40m/s, and the drag reduction rate is 3.84%.

The following is a preliminary feasibility assessment of the scheme with drag reduction effect:

(1) Locomotive blowing, this scheme is equivalent to increasing the traction of the train, bringing the speed in the opposite direction of the train wind so that it can reduce drag. In the real operating environment of the train, there will be many aspects of disturbance; in engineering, blowing is difficult to achieve, so it is not recommended to choose this scheme.

(2) Air blowing at the top of the nose tip at the rear of the car; this scheme is equivalent to a certain interference to the separation flow of the tail of the train so that drag reduction can be reduced. At this time, the maximum drag reduction speed is 40m/s, and the blower power is not too large to achieve this blowing speed in engineering. The drag reduction rate at this time is 3.84%, so the final drag reduction effect will be very obvious. Therefore, this scheme is recommended.

The above feasible evaluation shows that the drag reduction scheme of the upper part of the nose tip at the rear position is feasible. The blowing speed is 40m/s, and the drag reduction rate is 3.84%.

3.5. Research on drag reduction of locally arranged turbulence vortex generator

In this paper, the eddy current generator is applied to the rear of the train to reduce the resistance of the train. The installation position of the eddy current generator is determined by analyzing 3.4 sections. The model after the installation of the eddy current generator is shown in Figure 5. Through numerical calculation, the drag reduction size of the eddy current generator and the corresponding drag reduction rate is determined, as shown in Table 2.

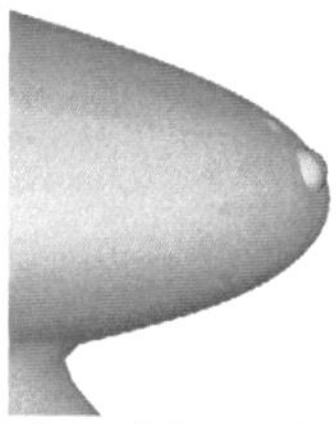

FIG. 5 Calculation model of the train with spherical spoiler

Table 2 Drag reduction scheme and drag reduction rate of eddy current generator

Scheme No.	Location of spoilers	Spoiler size (mm)	Original drag coefficient	Drag coefficient (plus spoiler)	Drag reduction efficiency
1	up	10	0.3650773	0.3540744	3.01%
5	up	50	0.3653259	0.3540764	3.08%
10	up	100	0.36567929	0.3541747	3.15%
11	down	10	0.36421115	0.3532408	3.01%
15	down	50	0.36436566	0.3532282	3.06%
20	down	100	0.36465071	0.3532351	3.13%
21	Up、down	10	0.36546787	0.354385	3.03%

25	Up、down	50	0.36574962	0.3543747	3.11%
30	Up、down	100	0.36630683	0.3544074	3.25%

As can be seen from Table 2, all 30 schemes have a drag reduction effect. The larger the spoiler size, the better the drag reduction effect, and the best drag reduction effect is when the spoiler is installed at both locations.

From the perspective of engineering application, since the train runs in both directions, the spoiler should be designed to be pulled out when needed. This drag reduction does not require additional power, so it is recommended to use.

3.6. Summary of the drag reduction scheme

Now the four optimization schemes are summarized, as shown in Table 3. According to the analysis in each section of Table 3 and Section 3, the bionic groove, eddy current generator, and blow suction drag reduction scheme can be used for train drag reduction.

Table 3 Summary of the drag reduction scheme

Scheme Name	Content of Solution	Drag reduction efficiency
Bionic groove	The train is partially equipped with bionic grooves	2.53%
All-inclusive outside windshield	Fill the connecting part of the train with soft material	8.17%
Blow and inhale	Partial air blowing at the rear of the train	3.84%
Eddy generator	A spoiler is placed at the rear of the train	3.25%

4. Conclusion

In this paper, a high-speed EMU model is constructed to simulate and analyze the outflow field condition of the train running on the open line. Compared with the experimental values, it can be seen that the calculation model and calculation method adopted in this paper are accurate and reliable. Through a detailed study of the pressure distribution on the surface of the car body and the flow of air around the train, an improvement scheme is proposed for the adverse areas, such as the high-pressure area and the flow area, that cause the overall resistance of the train to rise. Numerical simulation is carried out on the improvement schemes one by one. Through comparison, an optimization scheme with a better drag reduction effect is selected. The optimization scheme effectively reduces the overall resistance of the train. The following conclusions can be drawn from the research:

(1) When the original vehicle model runs at low speed, the airflow condition is relatively stable, the flow around the vehicle is less, and there is no large range of red high-pressure zone on the surface of the vehicle body. However, at high speed, the bumps on the surface of the outside of the vehicle seriously hinder the flow. There is a stagnation point on the windward side, and the scope of the red high-pressure zone increases. On the leeward side, airflow is separated, and a negative pressure zone is formed. The increase of front and rear pressure difference leads to the increase of resistance of the whole vehicle. The flow around the train surface with clearance is serious, which affects the patency of the whole flow field and increases the resistance value of the vehicle.

(2) In view of the modification of the large gap in the connecting position between the vehicle bodies, namely the inner windshield area, the whole outsourced windshield was added. The cost was high, but it could achieve no gap connection between the vehicle bodies and maintain good air fluency. After installation, the resistance of the whole vehicle was reduced by about 8.17%.

(3) In view of the large body of the train, the bionic groove is proposed to reduce the drag coefficient of the surface friction. By selecting the groove with the best size, the drag reduction effect is obtained, and the drag reduction can be about 2.53%

(4) For the separation area at the rear of the train, a local drag reduction scheme is proposed. Through the calculation of 49 schemes, the maximum drag reduction rate is 3.84%.

(5) For the separation area at the rear of the train, the local arrangement of the eddy current generator drag reduction scheme is proposed. By calculating 30 schemes, the maximum drag reduction rate is 3.25%.

References

[1] STH J, KRAJNOVIS S. A Study of the Aerodynamics of a Generic Container Freight Wagon Using Large-eddy Simulation [J]. Journal of Fluids and Structures, 2014, 44: 31-51.

[2] MUNOZ-PANIAGUA J, GARCIAJ. Aerodynamic Drag Optimization of a High-speed Train [J]. Journal of Wind Engineering and Industrial Aerodynamics, 2020, 204: 104215.

[3] DONG T Y, LIANG X F, KARJNOVICS, et al. Effect of Simplifying Bogies on Surrounding Flow and Aerodynamic Forces [J]. Journal of Wind Engineering & Industrial Aerodynamics, 2019, 191: 170-182.

[4] CHEN G, LI X B, LIU Z, et al. Dynamic Analysis of the Effect of Nose Length on Train Aerodynamic Performance [J]. Journal of Wind Engineering and Industrial Aerodynamics, 2019, 184: 198-208.

[5] CHOI J K, KIM K H. Effects of Nose Shape and Tunnel Cross-sectional Area on Aerodynamic Drag of Train Traveling in Tunnels [J]. Tunneling and Underground Space Technology incorporating Trenchless Technology Research, 2014, 41:62-73.

[6] GAO G J, LIF, HE K, et al. Investigation of Bogie Position on the Aerodynamic Drag and Near Wake Structure of a High-speed Train [J]. Journal of Wind Engineering & Industrial Aerodynamics, 2019, 185: 41-53.

Properties on Generalized Perron Complements of Inverse N_0-matrices

Qin Zhong [1], Chunyan Zhao [1], Gufang Mou [2], Ling Li [3*]

[1] Department of Mathematics, Sichuan University Jinjiang College, Meishan 620860, China

[2] College of Applied Mathematics, Chengdu University of Information Technology, Chengdu 610225, China

[3*] School of Big Data and Artificial Intelligence, Chengdu Technological University, Chengdu 611730, China

[*] Corresponding author E-mail: lynn910309@126.com

Abstract: For an irreducible and nonpositive matrix, we present the concepts of the Perron complement matrix and the generalized Perron complement matrix. An inequality that relates the generalized Perron complement matrices of inverse N_0-matrices is derived. In addition, we obtain the quotient formula of inverse N_0-matrices based on the quotient formula for the Schur complements.

1. Introduction

Let $B=(b_{ij})\in R^{n\times n}$. If $b_{ij}\geq 0$, then B is a nonnegative matrix, written as $B\geq 0$. If $b_{ij}\leq 0(i\neq j)$, then B is a Z-matrix.

The Z-matrix B is further referred to be an M-matrix if $B^{-1}\geq 0$. The properties and characterizations of M-matrices have been extensively explored by many experts and scholars.

In [1], Johnson defined the class of N_0-matrices as follows:

$$B=tI-P,\ P\geq 0, \tag{1}$$

where $\lambda\leq t<\rho(P)$ and λ is the maximal Perron root of all principal submatrices of a nonnegative matrix P with the dimension $n-1$.

The class of N_0-matrices has a close relationship with M-matrices. Some important results on the class of N_0-matrices can be found in [2]. Furthermore, there exists great interest in inverse N_0-matrices, i.e., whose inverse is an N_0-matrix. This paper is to study the properties of the inverse N_0-matrices.

Let $\langle n\rangle=\{1,2,\cdots,n\}$. α is a nonempty subset of $\langle n\rangle$, $\alpha\cup\beta=\langle n\rangle$. The elements of subsets α,β are strictly increasing integers. The submatrix of the matrix B whose rows are determined by α and columns are determined by β is denoted by $B[\alpha,\beta]$. For brevity, let us use $B[\alpha]$ to represent $B[\alpha,\alpha]$.

Assuming that the matrix B is partitioned in the pattern:

$$B=\begin{pmatrix} B[\alpha] & B[\alpha,\beta] \\ B[\beta,\alpha] & B[\beta] \end{pmatrix}, \tag{2}$$

then the Schur complement matrix of the non-singular submatrix $B[\beta]$ in B is the matrix:

$$B/B[\beta]=B[\alpha]-B[\alpha,\beta](B[\beta])^{-1}B[\beta,\alpha]. \tag{3}$$

In 1917, the definition of Schur complement was first introduced by Schur, which greatly promoted the development of mathematics, especially as an effective tool to deal with large-scale matrix calculation, and was widely used in matrix theory, statistics, applied mathematics, and other fields [3-4]. The Schur complement is crucial for determining norms, eigenvalues, singular value inequalities, and control inequalities in addition to matrices, determinants, traces, and other mathematical constructs. Since the late 1960s, there has been a significant amount of study on Schur complements of *M*-matrices and totally nonnegative matrices.

In [5], for the matrix B with nonnegative elements, the Perron complement matrix idea was introduced as follows:

$$P(B/B[\beta])=B[\alpha]+B[\alpha,\beta](\rho(B)I-B[\beta])^{-1}B[\beta,\alpha]. \tag{4}$$

The maximum eigenvalue and irreducibility of the matrix $P(B/B[\beta])$ were also explored and some interesting findings were studied. The concept of Perron complement was further generalized by Neumann [6]. A lot of work has been done on the Perron complements [7-8]. Our research is based on the observation that an N_0-matrix's inverse is nonpositive and irreducible. In this paper, for a nonpositive matrix, we will make a slight extension of the notation of the Perron complement.

Let B be a partitioned inverse N_0-matrix as (2). The Perron complement of the submatrix $B[\beta]$ in B is represented by the following formula:

$$P(B/B[\beta])=B[\alpha]-B[\alpha,\beta](\rho(B)I+B[\beta])^{-1}B[\beta,\alpha]. \tag{5}$$

(5) can be well-defined due to $\rho(B[\beta])<\rho(B)$.

If we replace $\rho(B)$ in (5) with t, we will get

$$B[\alpha]-B[\alpha,\beta](tI+B[\beta])^{-1}B[\beta,\alpha], \tag{6}$$

where $t>\rho(B[\beta])$. The above expression is defined to be the generalized Perron complement of the submatrix $B[\beta]$ in B at t and is denoted by $P_t(B/B[\beta])$.

In this paper, we continue to look into the characteristics of the generalized Perron complements of inverse N_0-matrices in light of the interesting applications of the Perron complements.

2. Main results

We start by introducing the subsequent lemmas.

Lemma 1 [9] If the matrices E,F are non-singular *M*-matrices such that $E\geq F$, then $0\leq E^{-1}\leq F^{-1}$.

Lemma 2 [1] Let the inverse of the matrix E be a *Z*-matrix, if $E\leq 0$ and is irreducible, then E is an inverse N_0-matrix.

For convenience, we will denote $P_t(B/B[\beta])$ by $P_t(B/\beta)$ and $B/B[\beta]$ by B/β, respectively.

In general, the generalized Perron complement is a nonlinear operation on matrices, that is

$$P_t(A+B/\beta)\neq P_t(A/\beta)+P_t(B/\beta).$$

But, for the class of inverse N_0-matrices, the following inequality on the generalized Perron complements holds.

Theorem 3 Let $n \times n$ matrices A, B be inverse N_0-matrices. If $\varnothing \neq \beta \subset \langle n \rangle$, $\alpha \cup \beta = \langle n \rangle$, then the following inequality holds:

$$P_t(A+B/\beta) \leq \frac{P_t(A/\beta)+P_t(B/\beta)}{2}, \tag{7}$$

where $t > \rho(A[\beta]+B[\beta])$.

Proof Assuming that the matrices A, B are symmetrically permuted to the block forms:

$$A = \begin{pmatrix} A[\alpha] & A[\alpha,\beta] \\ A[\beta,\alpha] & A[\beta] \end{pmatrix} = \begin{pmatrix} A_{11} & A_{12} \\ A_{21} & A_{22} \end{pmatrix},$$

and

$$B = \begin{pmatrix} B[\alpha] & B[\alpha,\beta] \\ B[\beta,\alpha] & B[\beta] \end{pmatrix} = \begin{pmatrix} B_{11} & B_{12} \\ B_{21} & B_{22} \end{pmatrix}.$$

By Lemma 2, the matrices A, B are nonpositive and irreducible. We have

$$A+B = \begin{pmatrix} A_{11}+B_{11} & A_{12}+B_{12} \\ A_{21}+B_{21} & A_{22}+B_{22} \end{pmatrix}.$$

So, the generalized Perron complement of the submatrix $A_{22}+B_{22}$ in $A+B$ is

$$P_t(A+B/A_{22}+B_{22}) = (A_{11}+B_{11}) - (A_{12}+B_{12})(tI+A_{22}+B_{22})^{-1}(A_{21}+B_{21})$$

and the generalized Perron complement of the submatrix A_{22} is

$$P_t(A/A_{22}) = A_{11} - A_{12}(tI+A_{22})^{-1}A_{21}.$$

Noting that

$$t > \rho(A_{22}+B_{22}) \geq \rho(A_{22}),$$

and

$$t > \rho(A_{22}+B_{22}) \geq \rho(B_{22}),$$

then the generalized Perron complements in (7) are well-defined. Because

$$tI + A_{22} + B_{22} \leq tI + A_{22},$$

based on Lemma 1, we have

$$(tI+A_{22}+B_{22})^{-1} \geq (tI+A_{22})^{-1} \geq 0.$$

Observing that

$$-(A_{12}+B_{12}) \geq -A_{12} \geq 0,$$

and

$$-(A_{21}+B_{21}) \geq -A_{21} \geq 0,$$

therefore, we get

$$(A_{12}+B_{12})(tI+A_{22}+B_{22})^{-1}(A_{21}+B_{21}) \geq A_{12}(tI+A_{22})^{-1}A_{21}.$$

Taking into account $A_{11}+B_{11} \leq A_{11}$, by definition, we gain

$$P_t(A+B/A_{22}+B_{22}) \leq P_t(A/A_{22}). \tag{8}$$

In the same way, we get

$$P_t\left(A+B/A_{22}+B_{22}\right)\le P_t\left(B/B_{22}\right). \tag{9}$$

Combining (8) with (9), we obtain the desired result.

For obtaining the second conclusion of this article, we need to recall the famous quotient formula of Schur complements [8].

If $\varnothing \ne \alpha_1,\alpha_2 \subset \beta \subset \langle n\rangle$, with $\alpha_1 \bigcup \alpha_2 = \beta$ and $\alpha_1 \bigcap \alpha_2 = \varnothing$, then

$$A/\beta = \left(A/\alpha_1\right)/\alpha_2. \tag{10}$$

The same consequence applies to the inverse N_0-matrices.

Theorem 4 Let the $n\times n$ matrix A be an inverse N_0-matrix. If $\varnothing \ne \alpha_1,\alpha_2 \subset \beta \subset \langle n\rangle$ with $\alpha_1 \bigcup \alpha_2 = \beta$ and $\alpha_1 \bigcap \alpha_2 = \varnothing$, then for any $t\in\left(\rho\left(A[\beta]\right),\infty\right)$, the identity holds:

$$P_t\left(A/\beta\right)=P_t\left(P_t\left(A/\alpha_1\right)/\alpha_2\right). \tag{11}$$

Proof Recalling that since A is nonpositive and irreducible, we have

$$t>\rho\left(A[\beta]\right)>\rho\left(A[\alpha_1]\right).$$

This ensures that the Perron complement $P_t\left(A/\alpha_1\right)$ in (11) can be well-defined. Without loss of generality, let A be symmetrically permuted to the block form:

$$A=\begin{pmatrix} A[\alpha] & A[\alpha,\beta] \\ A[\beta,\alpha] & A[\beta]\end{pmatrix}=\begin{pmatrix} A_{11} & A_{12} \\ A_{21} & A_{22}\end{pmatrix}, \tag{12}$$

where $\alpha=\langle n\rangle \setminus \beta$. For the blocking inverse N_0-matrix A, $P_t\left(A/\beta\right)$ is the matrix:

$$A_{11}-A_{12}\left(tI+A_{22}\right)^{-1}A_{21}. \tag{13}$$

Considering the following matrix

$$tI+A=\begin{pmatrix} tI+A_{11} & A_{12} \\ A_{21} & tI+A_{22}\end{pmatrix},$$

by the definition of the Schur complement, we have

$$\begin{aligned}\left(tI+A\right)/\beta &= tI+A_{11}-A_{12}\left(tI+A_{22}\right)^{-1}A_{21} \\ &= tI+P_t\left(A/\beta\right).\end{aligned}$$

We thus obtain

$$P_t\left(A/\beta\right)=\left(tI+A\right)/\beta-tI.$$

On the other hand, let

$$G=P_t\left(A/\alpha_1\right)=\left(tI+A\right)/\alpha_1-tI.$$

Computing the right-hand side of (11), we acquire

$$\begin{aligned}P_t\left(P_t\left(A/\alpha_1\right)/\alpha_2\right) &= P_t\left(G/\alpha_2\right) \\ &= \left(tI+G\right)/\alpha_2-tI \\ &= \left(\left(tI+A\right)/\alpha_1\right)/\alpha_2-tI.\end{aligned}$$

According to (10), we get

$$\left(\left(tI+A\right)/\alpha_1\right)/\alpha_2=\left(tI+A\right)/\beta.$$

Therefore, (11) follows immediately.

In particular, the following result can be obtained by setting $t=\rho\left(A\right)$ in Theorem 4.

Corollary 5 Let the $n \times n$ matrix A be an inverse N_0-matrix, if $\varnothing \neq \alpha_1, \alpha_2 \subset \beta \subset \langle n \rangle$ with $\alpha_1 \bigcup \alpha_2 = \beta$ and $\alpha_1 \bigcap \alpha_2 = \varnothing$, then the identity holds:

$$P(A/\beta) = P\left(P(A/\alpha_1)/\alpha_2\right).$$

3. Conclusion

The generalized Perron complement matrices and Perron complement matrices of a nonpositive irreducible matrix are defined in this paper. Our work is based on the fact that inverse N_0-matrices are nonpositive and irreducible. An inequality that relates the generalized Perron complements of the class of inverse N_0-matrices is derived. The quotient formula of generalized Perron complements, which is analogous to the quotient formula of Schur complements, is also obtained for the inverse N_0-matrices. In the future, we will go over the characteristics of generalized Perron complement matrices of inverse N_0-matrices in more detail and demonstrate that they are still inverse N_0-matrices.

Acknowledgments

This research was funded by the Chengdu University of Information Technology's Scientific Research Project (KYTZ202185), and the Education Department of Sichuan Province's Natural Science Research Project (18ZB0364).

References

[1] Johnson G A. (1982) A generalization of *N*-matrices. Linear Algebra Appl., 48: 201–217.
[2] Johnson G A. (1985) Inverse N_0-matrices. Linear Algebra Appl., 64: 215–222.
[3] Haynsworth E V. (1970) Applications of an inequality for the Schur complement. Proceedings of the American Mathematical Society, 24(3): 512-516.
[4] Zhang F Z. (2000) Schur complements and matrix inequalities in the Löwner ordering. Linear Alg. Appl., 321: 399–410.
[5] Meyer C D. (1989) Uncoupling the Perron eigenvector problem. Linear Algebra Appl., 114/115: 69–94.
[6] Neumann M. (2000) Inverses of Perron complements of inverse *M*-matrices. Linear Algebra Appl., 313: 163-171.
[7] Ren Z G, Huang T Z, Cheng X Y. (2006) A note on generalized Perron complements of *Z*-matrices. The Electronic Journal of Linear Algebra, 15: 8-13.
[8] Fallat S M, Neumann M. (2001) On Perron complements of totally nonnegative matrices. Linear Algebra Appl., 327: 85–94.
[9] Fiedler M, Pták V. (1962) On matrices with non-positive off diagonal elements and positive principal minors. Czechoslovak Math. J., 12: 382-400.

Journal of Physics: Conference Series 2455 (2023) 012005 doi:10.1088/1742-6596/2455/1/012005

Numerical Simulation on Structural Integrity of Grain with Large Increased Area Ratio for Solid Rocket Motor

S X Wang*, J Q Bai, Y X Zuo, X W Qiang, H L Wang and Q X Yan

The 713 Research Institute of CSSC, 126 Jingguang Road, Zhengzhou, 450015, China

819523567@qq.com

Abstract. To evaluate the structural integrity of a type of solid rocket motor grain with a largely increased area ratio using free-loading modified double-base grain, the Schapery integral nonlinear viscoelastic constitutive equation with damage is derived, and the numerical simulation model of solid rocket motor grain is constructed. According to the mission profile of the grain during the equipment period, the numerical simulation research on the mechanical response characteristics of the grain under three typical working conditions, namely, platform vibration, underwater explosion impact, and ignition pressurization during the launch is carried out. The simulation results show that under the three typical working conditions, the grain will not produce structural damage and defects, and the mechanical response of the grain under the platform vibration and explosion impact load is far less than that under the launch ignition pressurization. The ignition pressurization should be considered in the grain structural integrity design of this type or similar structure solid rocket motor.

1. Introduction

The solid rocket motor can be used as the launching power source of missile weapons. During its mission, the grain is subjected to external loads such as platform vibration, explosion impact, and ignition pressurization. Relevant studies found that [1-3], under the action of external loads, internal damage is easy to occur and gradually accumulates, affecting the structural integrity of the grain, and thus affecting the combustion process of a solid engine and the ejection take-off process of missile weapons. Therefore, it is very important to study the structural integrity of solid rocket motor grain.

At present, by establishing the solid rocket motor grain structure model, and using the finite element method, to calculate the structural integrity of the grain and obtain the positions and links prone to damage, because of its low research cost, high calculation accuracy, and high analysis efficiency, it has become an important means to carry out the structural integrity research of solid rocket motor grain at home and abroad. Tunc B et al. [4,5] deduced and established the viscoelastic constitutive model with damage and expanded the model to a three-dimensional form. Using the finite element simulation software combined with the secondarily developed subprogram, the structural integrity simulation calculation of solid propellant grain was carried out. The numerical simulation and test results were in good agreement, which verified the reliability and feasibility of this numerical calculation method applied to the structural integrity study of grain. Meng H L [6] established the nonlinear viscoelastic constitutive equation with cumulative damage of the modified double-base propellant materials, compiled the constitutive model subprogram, and carried out numerical calculation of the structural integrity of the modified double-base propellant grain with ABAQUS software, and analyzed the influence of pressure load, star grain geometric parameters and cladding layer materials on the structural integrity of the grain. Deng K Q et al. [7] used the three-dimensional finite element method to calculate the mechanical response of a free-loading solid rocket motor during the whole process from solidification cooling to low-temperature ignition, which can provide a reference for the simulation study of the structural integrity of the free-loading solid rocket motor grain. Gao F L [8] carried out a numerical simulation study on the structural integrity of the grain against the background of explosion during the low-temperature ignition test of a solid rocket motor, guided the optimization of the design improvement of the grain structure, and solved the failure of the low-temperature ignition of the motor. Wang H L [9] applied the viscoelastic plastic constitutive equation with damage of modified double-base propellant materials to the finite element simulation calculation, carried out simulation research on star grain under ignition pressure load and acceleration load, and obtained the weak parts of the stress and strain of the grain. Wang C F [10] used Python language to conduct the parametric modeling of the solid rocket motor, carried out the structural integrity simulation research of large aspect ratio grain, and proposed the optimization strategy for the structural integrity of grain with the framework structure and stress relief cover structure according to the simulation results. At present, the relevant scholars at home and abroad have carried out a lot of work on the simulation study of the structural integrity of the grain, but the research only focuses on the specific charging structure and corresponding working conditions, such as low-temperature ignition, high overload, etc., and rare research studies the structural integrity of the free-loading motor grain under vibration and explosion impact load. Moreover, there is no simulation study on the structural integrity of the solid rocket grain with a largely increased area ratio. Therefore, it is particularly necessary to study the structural integrity simulation of free-loading large increased area ratio solid motor to double-base grain, and relevant work should be carried out urgently.

In this paper, the structural integrity of the free-loading modified double-base grain of the largely increased area ratio solid motor is simulated and analyzed. The nonlinear viscoelastic constitutive equation describing the mechanical properties of the modified double-base grain is adopted, the finite element analysis software ABAQUS and the secondary development of the UMAT subroutine are used, in combination with the grain task profile, considering the platform vibration, underwater explosion impact and ignition impact loads that the grain is subjected to during the task, the finite element simulation calculation is carried out, the mechanical response of the grain is obtained, and the impact of external loads on the structural integrity of the grain is analyzed to check the design rationality of grain structure and buffer protection measures and provide a reference for the subsequent design optimization and improvement of grain structure.

2. Nonlinear viscoelastic constitutive equation with damage

The mechanical behavior of modified double-base grain materials shows nonlinear characteristics, belonging to the category of viscoelastic mechanics. To accurately characterize the mechanical response of the modified double-base propellant with free loading, it is necessary to establish the constitutive equation to accurately describe the stress and strain of the modified double-base propellant. In this paper, the Schapery integral nonlinear viscoelastic constitutive equation is adopted, the specific form is as follows:

$$\sigma = E_R(1-D)\varepsilon^R \tag{1}$$

$$\varepsilon^R = \frac{1}{E_R}\int_0^t E(t-\tau)\frac{d\varepsilon}{d\tau}d\tau \tag{2}$$

As in Equations (1) and (2), σ is the stress, E_R is the reference elastic modulus, D is the damage variable, ε^R represents the pseudo strain, and $E(t)$ is the relaxation modulus of the material, which can be obtained through the tensile relaxation test.

The above nonlinear viscoelastic constitutive equation with damage is one-dimensional, while the UMAT subroutine in the finite element simulation software needs to be based on the three-dimensional constitutive equation for secondary development. Now, the incremental form of the three-dimensional viscoelastic constitutive equation is derived by a numerical expansion of Equation (1), and the stress updating expression and tangent modulus matrix (Jacobian matrix) is obtained.

The pseudo strain ε_{ij}^R is composed of partial pseudo strain e_{ij}^R and volumetric pseudo strain ε_{kk}^R in the following form:

$$e_{ij}^R = \int_0^t 2G(t-\tau)\frac{de_{ij}}{d\tau}d\tau \tag{3}$$

$$\varepsilon_{kk}^R = \int_0^t 3K(t-\tau)\frac{d\varepsilon_{kk}}{d\tau}d\tau \tag{4}$$

So the stress tensor

$$\sigma_{ij} = E_R(1-D)\varepsilon_{ij}^R = E_R(1-D)\left(e_{ij}^R + \frac{1}{3}\delta_{ij}\varepsilon_{kk}^R\right) \tag{5}$$

At $t+\Delta t$, the pseudo strain $\varepsilon_{ij}^R(t+\Delta t)$

$$\varepsilon_{ij}^R(t+\Delta t) = \varepsilon_{ij}^R(t) + \Delta\varepsilon_{ij}^R(t+\Delta t) = \varepsilon_{ij}^R(t) + \Delta e_{ij}^R(t+\Delta t) + \frac{1}{3}\delta_{ij}\Delta\varepsilon_{kk}^R(t+\Delta t) \tag{6}$$

The updated stress tensor

$$\sigma_{ij}(t+\Delta t) = E_R\left[1-D(t+\Delta t)\right]\varepsilon_{ij}^R(t+\Delta t) \tag{7}$$

For viscoelastic materials, the uniform tangent modulus

$$M_{ijkl}^T(t+\Delta t) = \frac{\partial\sigma_{ij}(t+\Delta t)}{\partial\varepsilon_{kl}(t+\Delta t)} \tag{8}$$

Then the tangent modulus applied to three dimensions can be expressed as

$$M_{ijkl}^T(t+\Delta t) = \frac{\partial\sigma_{ij}(t+\Delta t)\partial\varepsilon_{ij}^R(t+\Delta t)}{\partial\varepsilon_{ij}^R(t+\Delta t)\partial\varepsilon_{kl}(t+\Delta t)} \tag{9}$$

3. UMAT subroutine operation flow and verification

3.1. UMAT subroutine operation flow

According to the above discrete method of the three-dimensional viscoelastic constitutive model with damage, the UMAT subroutine of the modified double-base propellant material constitutive model is compiled. The calculation process and steps are as follows:

① Reading the material parameters and the damage variable $D(t)$ at time t, the stress $\sigma_{ij}(t)$, the strain $\varepsilon_{ij}(t)$, and the time increment Δt at the current moment which are passed into the UMAT subroutine by the interface of the main program software.

② Calculate the pseudo strain $\varepsilon_{ij}^R(t)$, the pseudo strain increment $\Delta\varepsilon_{ij}^R(t+\Delta t)$, and update the pseudo strain $\varepsilon_{ij}^R(t+\Delta t)$.

③ Update the damage variable $D(t+\Delta t)$ and judge whether the upper limit value D_{max} of damage is reached. If the upper limit value is not reached, the calculation will continue; otherwise, skip to step ⑥ and the calculation will be terminated.

④ Calculate the stress tensor $s_{ij}^{R}(t+\Delta t)$ and Jacobian increment matrix considering the damage variable.

⑤ Carry out the incremental iteration calculation, and judge whether the calculation converges. If not, update the time increment step to $\frac{\Delta t}{2}$, back to step ②, and perform the calculation again.

⑥ When the calculation converges, store the damage variable, stress, and strain, and this calculation ends.

After the calculation is completed, the calculation results will be fed back to the main program, waiting for the call of the next integration moment, so far as the call operation of the integration point subprogram is completed.

3.2. UMAT subroutine verification

To check the accuracy of the numerical simulation method and UMAT subroutine, the UMAT subroutine of the nonlinear viscoelastic constitutive model with damage is combined with the general nonlinear finite element simulation software to perform nonlinear finite element numerical simulation on the propellant sample, compare the simulation results with the sample test results, and modify the subroutine according to the coincidence of the results.

In this paper, uniaxial tensile test results of modified double-base propellant samples are used as a reference to verify the subroutine. According to GJB 770B-2005 "Test method of propellant", dumbbell-shaped samples are used for the uniaxial tensile test of propellants. The finite element simulation model and mesh division are shown in Figure 1. The two sides of the model are sample fixtures, and the middle is a sample, whose size and shape are consistent with the test requirements. To simulate the real stress state of the specimen during the tensile test, one fixture is subject to displacement constraint, and the other fixture is subject to constant velocity displacement load in the tensile direction so that the strain of the specimen in the simulation process is consistent with that in the tensile test process.

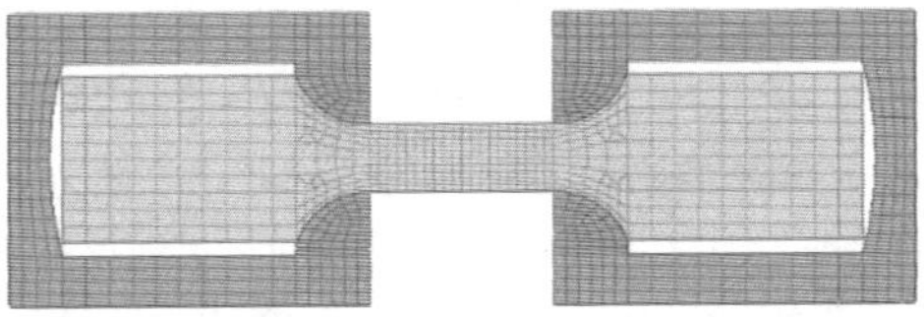

Figure 1. Simulation model and mesh division of the dumbbell sample.

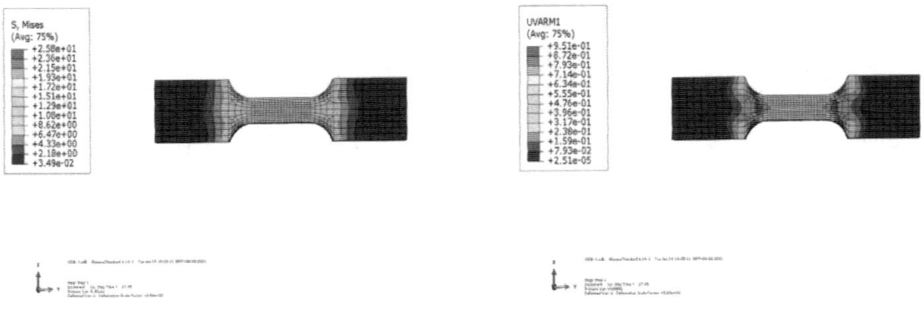

(a) Stress. (b) Strain.

Figure 2. Cloud chart of sample simulation results at a tensile rate of 50 mm/min.

Figure 2 shows the cloud chart of simulation results of the uniaxial tensile of a dumbbell-shaped sample of modified double-base propellant at the tensile rate of 50 mm/min. As can be seen from the figure, obvious contact stress and large local strain appeared on the contact edge between the fixture and the specimen. The standard distance section of the specimen is uniformly stretched and the stress-strain distribution is uniform, indicating that the numerical method and the subroutine are reasonable and feasible.

Point A, the middle part of the marking distance section of the specimen, is selected to extract the stress-strain curve of this point in the simulation process, and compared with the test curve. As shown in Figure 3, the numerical simulation results are in good agreement with the test results, with a relative error of less than 3%, indicating that the finite element method and the UMAT subroutine compiled are applicable to the numerical simulation of the modified double-base propellant with high accuracy.

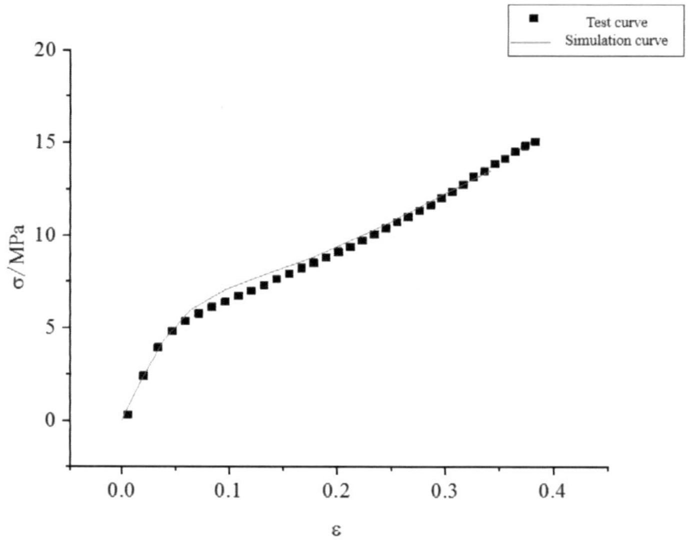

Figure 3. Comparison of uniaxial tensile test curve and simulation curve.

4. Finite element modeling

4.1. Geometric model and mesh generation

The object of study in this paper is a solid motor grain with a largely increased surface ratio, which is composed of several coated grains arranged in a specific sequence and free-loading combination. Each grain is provided with a baffle plate and an elastic washer at both ends to play the role of fixing and buffering. A single-coated grain is a cylinder with a length of about 600 mm, an outer diameter of about 100 mm, a through hole in the center, and a diameter of about 20 mm. The inside and outside surfaces burn at the same time during operation to provide initial speed for load takeoff. The loading state of the single grain is shown in Figure 4, and the boundary conditions and load constraints of each grain are consistent. Therefore, the single grain is selected as the simulation research object in this paper. To reduce the calculation amount, some structures are properly simplified during the modeling process.

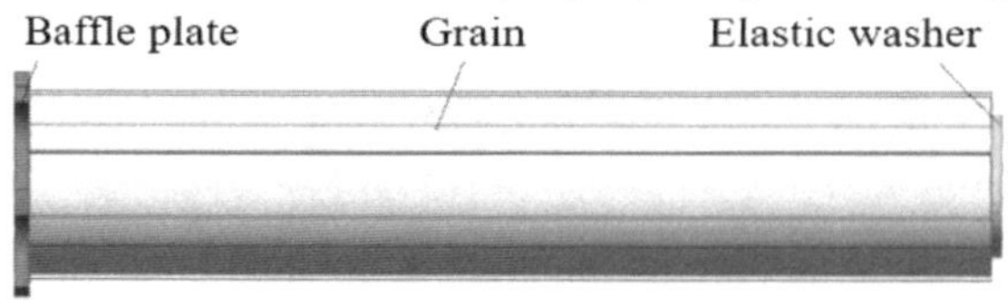

Figure 4. Simulation geometry model.

Hypermesh software is used to select hexahedral structured mesh to conduct three-dimensional modeling of the simulation geometric model and its complete entity. The element type is a three-dimensional eight-node entity unit C3D8. The mesh division is shown in Figure 5. The simulation model is divided into 72780 grid cells, and the mesh quality is checked, which is good. See Table 1 for detailed results.

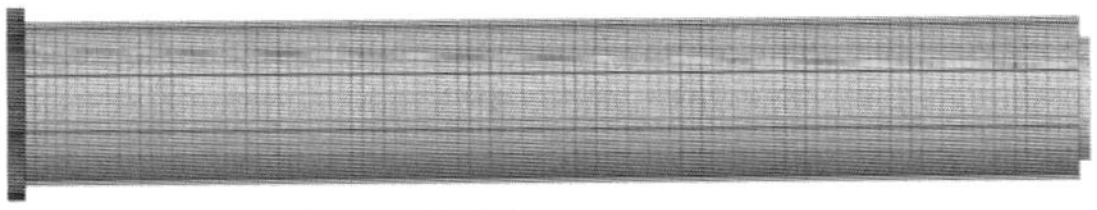

Figure 5. Mesh generation.

Table 1. Mesh quality check.

Mesh quality evaluation index	Requirements	Limit value
warpage	< 5°	0.09°
aspect	< 5	3.49
skew	< 60°	0
jacobian	> 0.7	0.75
tet collapse	Collapse or not	None
vol skew	Twist or not	None
vol AR	< 5	3.49

4.2. Material parameters

The simulated geometric model includes the baffle plate, the grain, the coating layer, and an elastic washer. The baffle plate is made of 921 steel, which belongs to the metal elastic material. The composition of the grain is modified double-base propellant, which belongs to viscoelastic material. The coating layer is nitrocellulose paint cloth, whose mechanical properties are similar to those of the modified double-base propellant, and the same material parameters are used in the simulation. The elastic washer is a sponge rubber sheet. The basic material parameter values of each part are shown in Table 2.

Table 2. Material parameters.

Part	Material	Density ($kg \cdot m^{-3}$)	Modulus (MPa)	Poisson's ratio	Thermal conductivity (W/m·K)	Specific heat (J/kg·K)
Baffle plate	921 steel	7890	206000	0.30	47	512.91
Grain	Modified double-base propellant	1690	/	0.495	0.418	1267
Coating layer	Nitrocellulose paint cloth	1690	/	0.495	0.418	1267
Elastic washer	Sponge rubber sheet	391	0.7905	0.40	/	/

The 921 steel and sponge rubber plate adopt the elastic material model provided by ABAQUS software, while the nonlinear viscoelastic constitutive model is adopted for modified double-base propellant and nitrocellulose paint cloth. The calculation support is provided by the UMAT subroutine, and the modulus is the tensile relaxation modulus. The relaxation modulus function of the sixth-order Prony series adopted in this paper is as follows:

$$E(t) = 31.89 + 10.89e^{-\frac{t}{0.01}} + 11.54e^{-\frac{t}{0.1}} + 12.81e^{-t} + 6.26e^{-\frac{t}{10}} + 24.16e^{-\frac{t}{100}} + 5.17e^{-\frac{t}{1000}} \quad (10)$$

5. Simulation results and analysis

The grain is subjected to various and complex loads during service. In this paper, three typical loads, platform vibration, underwater explosion impact, and ignition impact, are selected to carry out a numerical simulation analysis of the structural integrity of the grain.

5.1. Vibration condition

The subspace steady state dynamic analysis method is used to calculate the stress and strain response of the grain under vibration load. The basic idea of subspace steady state dynamic analysis is to extract the characteristic modes of undamped and symmetric systems, select appropriate feature vectors to form the characteristic mode subspace, project the steady state dynamic equations onto the characteristic mode subspace, and solve the steady state dynamic equations in the subspace by a direct method.

5.1.1. Boundary conditions and calculation conditions

The boundary conditions are set as follows: tie contact is used between the shield plate and the coated grain, between the coated grain and the elastic washer, and between the grain and the coating layer. Through the harmonic response analysis of ABAQUS, the elastic washer side and the baffle plate side were respectively restrained. The inner and outer surfaces of the grain are free boundaries.

The simulation model is an axisymmetric structure, which requires numerical analysis of vibration loads in both radial and axial directions. The applied displacement and acceleration load parameters are shown in Table 3.

Table 3. Applied parameters of vibration load.

Frequency	Parameter Index	Duration of vibration exposure	Remarks
1 Hz ~ 13.2 Hz	Displacement ≥ 1 mm	Perform 10 frequency sweep cycles	Sweep frequency at the rate of 1 oct/min
13.2 Hz ~ 24 Hz	Acceleration ≥ 7 m/s^2		

5.1.2. Simulation results and analysis

(1) Radial vibration load

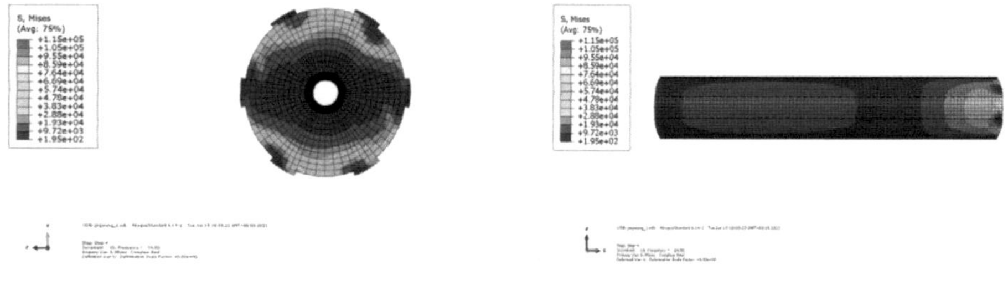

(a) Mises stress.

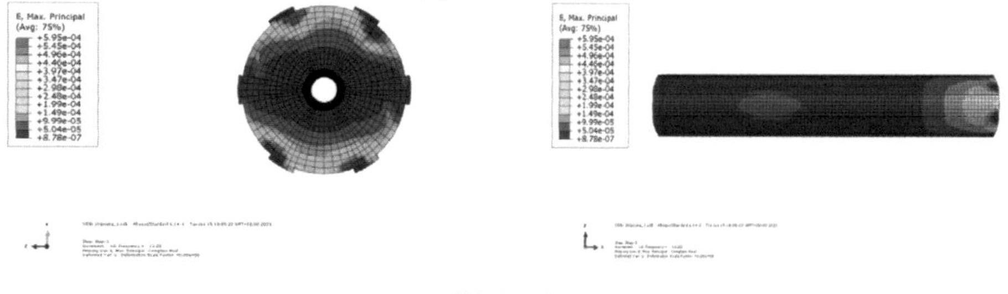

(b) Strain.

Figure 6. Stress-strain response of grain under radial vibration load.

The stress-strain response of the grain under radial vibration loading is shown in Figure 6. It can be seen from the cloud chart of the simulation results that the stress-strain response gradually increases along the radius direction from the inner hole to the outer surface, and the maximum stress-strain response appears at the end position and the position of the tooth root of the grain.

During the whole sweep frequency cycle along the radial vibration, the relationship curve between the maximum stress-strain response of the grain and the frequency is shown in Figure 7. The maximum response increases with the increase of frequency, the speed increases rapidly in the frequency range of 1 ~ 13.2 Hz and slows down in the frequency range of 13.2 ~ 24 Hz. The maximum stress in the whole process is 0.115 MPa, and the maximum strain is 0.0595 %.

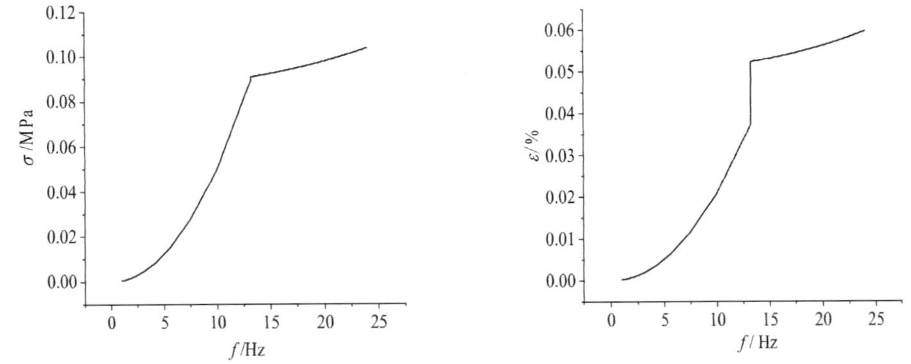

(a) Stress-frequency curve. (b) Strain-frequency curve.

Figure 7. Relation between maximum stress-strain response and frequency of grain under radial vibration load.

(2) Axial vibration load

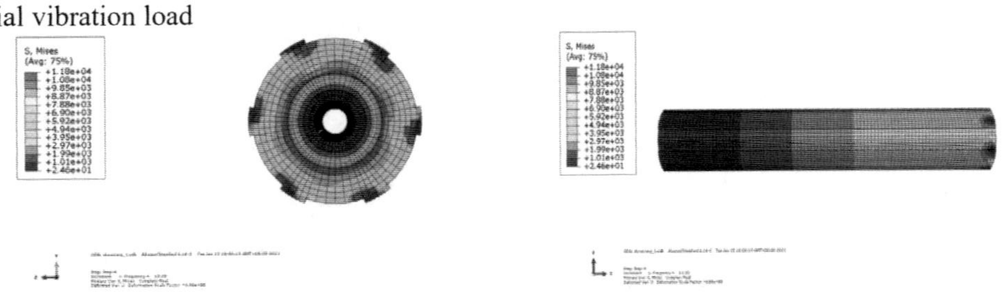

(a) Mises stress.

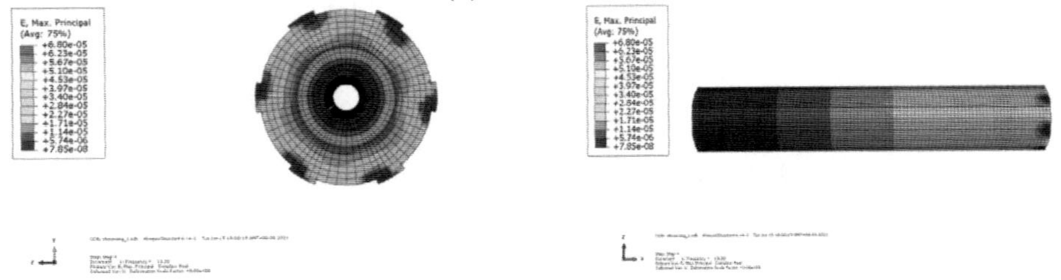

(b) Strain.

Figure 8. Stress-strain response of grain under axial vibration load.

Figure 8 shows the stress-strain response of the grain under axial vibration load. It can be seen from the cloud chart of simulation results that the stress-strain response increases gradually along the radius direction from the inner hole to the outer surface, and the maximum stress-strain response also appears at the end position and the root of the tooth of the grain. However, slightly different from the response results under radial vibration load, the stress and strain response at the tooth position of the grain is approximately similar to the position of the inner hole of the grain.

During the whole sweep frequency cycle along the axial vibration, the relationship curve between the maximum stress-strain response of the grain and the frequency is shown in Figure 9. Within the frequency range of 1 ~ 13.2 Hz, the maximum stress-strain response increases with the increase of the frequency, and within the frequency range of 13.2 ~ 24 Hz, the maximum response decreases slightly with the increase of the frequency. In the whole process, the maximum stress is 0.0118 MPa, and the maximum strain is 0.0068 %.

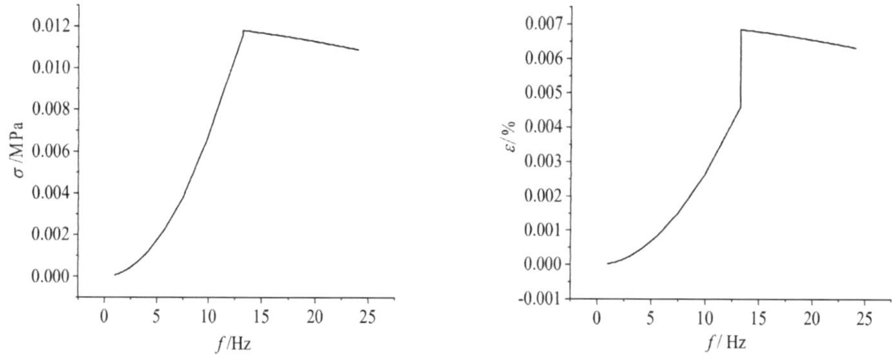

(a) Stress-frequency curve. (b) Strain-frequency curve.

Figure 9. Relation between maximum stress-strain response and frequency of grain under radial vibration load.

According to the results of the simulation calculation, the maximum response of the grain under axial vibration is smaller than that under radial vibration. This is because the grain is equipped with elastic washers along the axial direction, which can absorb energy during vibration and have a cushioning effect to reduce the load stimulation from axial vibration.

No matter the radial vibration or axial vibration, the maximum response value is evenly distributed at the end face and the root of the tooth of the grain. This is because the end face of the grain is in hard contact with the baffle plate and there is no transition interface. At the same time, the external surface profile of the grain is not smooth transition at the root of the tooth, which is easy to produce the phenomenon of stress concentration, which is the weak link of the grain structure. However, the maximum stress-strain response values are far less than the yield threshold of propellant materials, indicating that the grain under this state is in the elastic deformation range. Once the load is removed, the grain will return to its original state, so the platform vibration condition will not cause damage to the grain structure.

5.2. Impact condition

5.2.1. Boundary conditions and calculation conditions

The setting of the boundary conditions is the same as that of the vibration condition, which is calculated using the display dynamics analysis step of ABAQUS. The impact load parameters used in the calculation are shown in Table 4.

Table 4. The impact load parameters.

Peak acceleration	Waveform	Duration
300 m/s^2	Half sine wave	40 ms

5.2.2. Simulation results and analysis

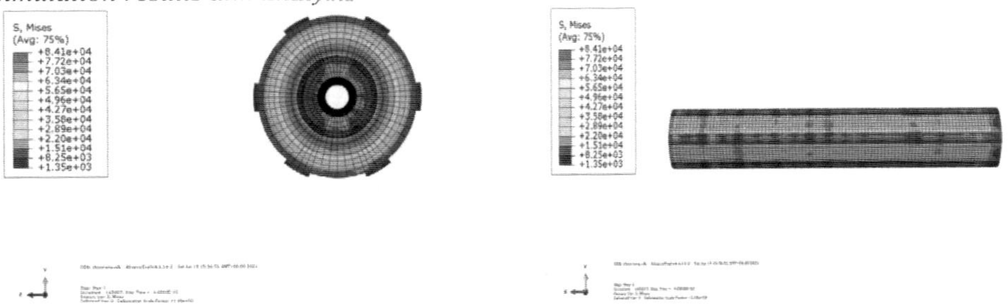

(a) Mises stress.

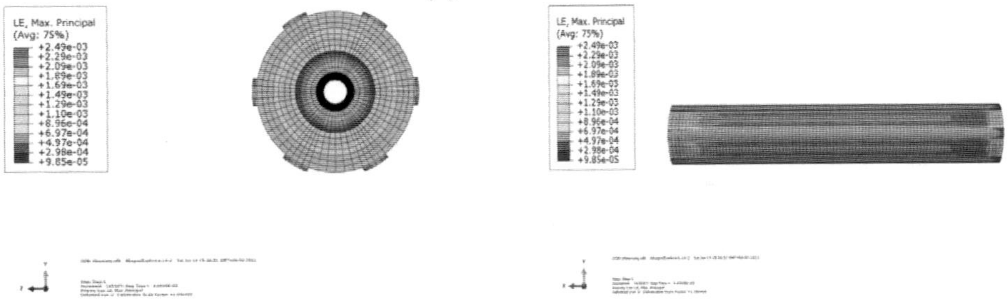

(b) Strain.

Figure 10. Stress-strain response of grain under impact load.

The cloud chart of stress-strain response results of a single grain under impact load is shown in Figure 10, as shown in Figure 10 (a), the Mises stress of the grain presented a circular distribution in the radial direction, and the stress circle increased first and then decreased from the inner hole to the outer surface of the grain. Along the axial direction, the stress presented strip distribution, and there is an obvious stress boundary between the position of grain tooth surface and tooth root, and the stress value at the circular surface of the tooth root is greater than that at the position of the tooth surface. On the whole grain, the maximum stress appears on the fixed end face of the contact side of the grain and the baffle plate, and the corresponding stress value was 0.082 MPa. The reason why the maximum stress appears here is that the two are in direct contact here without any buffer measures, which belongs to hard contact. When the grain is subjected to an external impact load, the elastic modulus of the baffle plate is much larger than the grain modulus, which is equivalent to the load being directly imposed on the end face of the grain, and the grain directly bears the impact effect.

As shown in Figure 10 (b), the strain of the grain under impact load is distributed circularly along the radial direction, first increasing and then decreasing from the inner hole to the outer surface of the grain. The strain is distributed along the axial strip. There is an obvious boundary between the position of the tooth surface and the tooth root, and the stress value at the circular surface of the tooth root is greater than that at the tooth surface. The maximum strain of the whole grain appears on the tooth root circular surface of the outer surface, and the corresponding strain value was 0.321 %.

It can be seen that the magnitude of stress-strain response under impact load is small, the grain deformation is still within the elastic range, and the grain can be restored to its original state after the external load is removed. Therefore, according to the simulation results, this impact condition will not cause damage to the grain.

5.3. Ignition impact condition

5.3.1. Boundary conditions and calculation conditions

The setting of boundary conditions: the initial compression displacement of 2 mm is applied to the end of the elastic washer to simulate the initial preloading state under the condition of grain assembly. The pressure load is applied to the inner surface, the outer surface, and both ends of the grain. According to the measured values of the motor pressure many times, the ignition impact load is determined to be 6 MPa. The other conditions are set the same as the vibration condition.

5.3.2. Simulation results and analysis

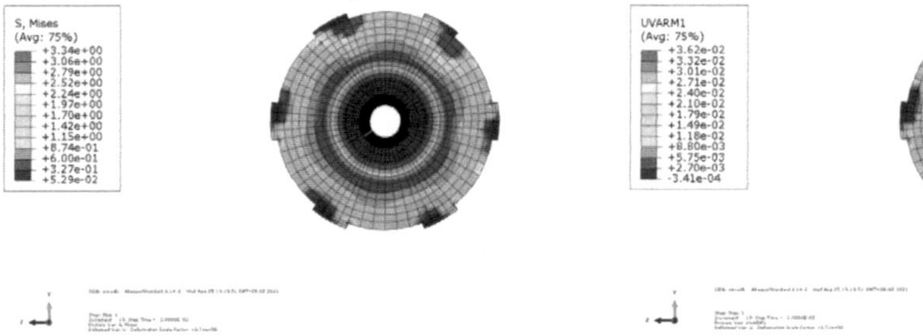

(a) Baffle plate end.

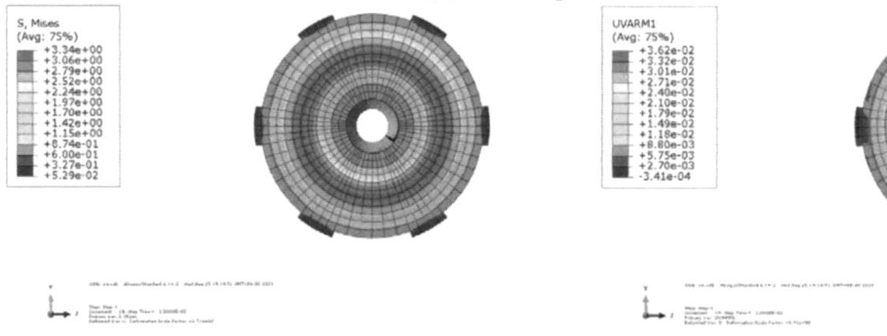

(b) Elastic washer end.

Figure 11. Stress-strain response of grain end face under ignition impact load.

Figure 11 shows the stress-strain response of the grain under the ignition impact load. It can be seen that the stress strain shows a circular distribution on the end result cloud chart. As shown in Figure 11 (a), on the end face at the contact side between the grain and the baffle plate, the value of stress-strain gradually increases from the inner hole to the outer surface, reaching the maximum at the root of the tooth. On the end face at the contact side between the grain and the elastic washer, as shown in Figure 11 (b), the stress-strain value increases first and then decreases from the inner hole to the outer surface along the radial direction. On the whole, the maximum Mises stress value of the grain is 3.34 MPa, and the maximum equivalent strain value of the grain is 3.62 %, both of which are less than the maximum tensile strength and maximum elongation of this propellant material measured by the test, indicating that the structural integrity of the grain is not damaged. Therefore, under the ignition impact load, the grain can burn according to the predetermined design burning surface, and will not produce defects or damage due to the ignition impact load.

6. Conclusion

In this paper, by modeling a certain grain and using the UMAT subroutine, the stress-strain response of the grain under the corresponding load excitation is numerically simulated according to the typical loads (platform vibration, underwater explosion impact, and ignition impact) that the grain is subjected to during its service. The results show that:

(1) The viscoelastic constitutive model with damage and the UMAT subroutine can accurately describe the mechanical response of modified double-base propellant materials.

(2) The stress-strain response of the grain under vibration load and impact load is far less than that in the process of ignition and pressurization. In comparison, the transient process of ignition and pressurization is the load condition that should be mainly considered in the study of the structural integrity of the grain.

(3) No matter the vibration, impact, or ignition pressure load, the grain will not produce damage and defects, the structural integrity will not be affected, and the grain can burn according to the set state.

References

[1] Bao F T and Hou X 2016 *Design of solid rocket motor* (Beijing: China Astronautic Publishing House)
[2] Gong J L, Li P, Zhang Z Z, Gong W J and Li W F 2021 Recent progress upon structural integrity analysis of solid rocket motor grain. *Science Technology and Engineering*. **21(2)** 434-441
[3] Wei X L, Zhou J H, Li H Y and Hu H M 2022 Research progress on structural integrity analysis of solid propellant charge. *Journal of Ordnance Equipment Engineering*. **43(01)** 19- 26
[4] Tunc B and Ozupek S 2016 Implementation and validation of a three dimensional damaging finite strain viscoelastic model. *International Journal of Solids and Structures*. **102(12)** 275- 285
[5] Tunc B and Ozupek S 2017 Constitutive modeling of solid propellants for three dimensional nonlinear finite element analysis. *Aerospace Science and Technology*. **69(8)** 290-297
[6] Meng H L 2011 Research on numerical simulation method of structure integrity analysis for modified double-base propellant grain (Nanjing: Nanjing University of Science and Technology)
[7] Deng K Q, Zhang L, Pang A M, Yu R, Yang L and Xin B B 2018 Analysis on structure integrity of a free loading solid propellant grains under ignition loading at low temperature. *Journal of Solid Rocket Technology*. **41(4)** 428-434
[8] Gao F L 2012 Research of structural integrity for a certain solid rocket motor grain (Changsha: National University of Defense Technology)
[9] Wang H L 2018 Research on visco-elastic-plastic constitutive model with damage of modified double-base propellant and its application (Nanjing: Nanjing University of Science and Technology)
[10] Wang C F 2018 Analysis on structural integrity of complex charge with large length-diameter ratio (Nanjing: Nanjing University of Science and Technology)

Journal of Physics: Conference Series **2455** (2023) 012006 doi:10.1088/1742-6596/2455/1/012006

Sliding bifurcations of a two DOF self-excited SD oscillator with Coulomb friction

Zhixin Li[1*], Zhiguo Li[2], Ke Zhang[1] and Jiaqi Lu[1]

[1] School of Mathematics and Physics, Hebei University of Engineering, Handan 056038 P. R. China

[2] School of Science, Hebei University of Technology, Tianjin 300401, P. R. China

*Corresponding author: lizhixin@hebeu.edu.cn

Abstract. In this study, on the basis of the SD (smooth and discontinuous) oscillator, we analyze the sliding bifurcations of a two DOF self-excited system driven by the classical conveyor belt friction. In order to elaborate the nonsmooth model named the self-excited SD oscillator, we model the friction of the conveyor belt of this system as Coulomb friction. Switching surfaces and sliding regions of this nonsmooth system are obtained based on the Filippov theory. The numerical simulations are performed to present the various scenarios of sliding bifurcation and chaotic attractors in the system. The results of this study give us an opportunity to understand clearly the bifurcation mechanism of the rich nonlinear frictional dynamics in machinery field.

1. Introduction

Generally speaking, the problem of friction-induced self-excited vibration has been the focus of many researchers in the field of practical mechanical engineering, i.e., chattering machine tools [1], brake [2], drill string [3], and others. The phenomenon of stick-slip motion is often encountered in the self-excited system with dry friction, which was investigated in seismology, especially since fault dynamics [Figure 1 (a)] was often modeled as the Burridge–Knopoff model[4], as shown in Figure. 1 (b), which describes the interaction between two crustal plates in a geologic fault structure as a chain of blocks coupled elastically to each other and to one of the plates, and suffering a frictional force from the surface of the another plate, such that they execute the stick-slip motion.

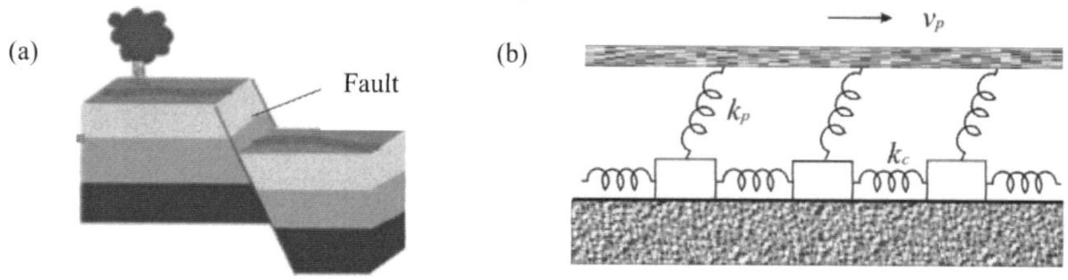

Figure 1. (a) Fault caused by earthquake; (b) the Burridge-Knopoff model.

In this study, in order to simplify the Burridge-Knopoff model, we analyze the sliding bifurcations of a two DOF system named after the self-excited SD oscillator [5], on the basis of the SD oscillator [6,7] and the conveyor belt. This system is characterized by multiple stick regions, multiple stick-slip phenomenoa, and coexistence of periodic motions. Dynamics behaviors such as multiple stick-slip chaotic motions [8], the global and local bifurcations [9], and stick-slip vibrations [10] of this system were analyzed.

2. Mechanical model

In this section, this self-excited mechanical system is sketched in Figure 2, which is composed of two

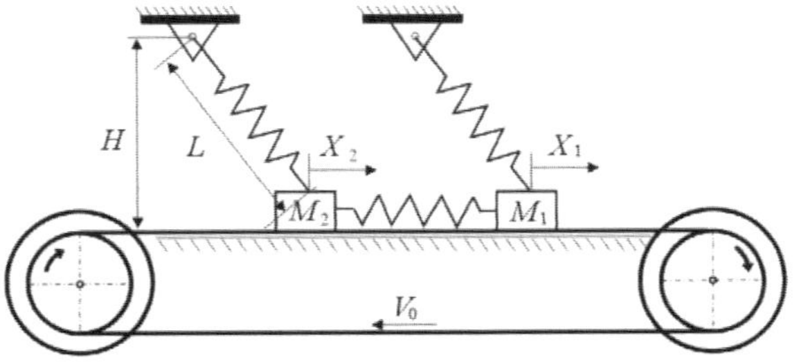

Figure 2. The schematic diagram of the two DOF self-excited SD oscillator.

blocks of mass M_1 and M_2, both supported by a indeformable conveyor belt in the case of a constant speed V_0, respectively connected to a fixed rack by an oblique linear spring of rigidity coefficient K_0. Two masses are connected together by an aclinic linear spring of rigidity coefficient K_c, which is able to resist both tension and compression. The contact area between the masses and the conveyor belt is considered rough so that the conveyor belt applies the frictional forces F_{s1} and F_{s2} on these two masses.

The motion equations of this system can be expressed in the following second-order differential equations

$$\begin{cases} M_1\ddot{X}_1 + K_0X_1\left(1-\dfrac{L}{\sqrt{X_1^2+H^2}}\right) + K_c(X_1-X_2) + F_{s1} = 0, \\ M_2\ddot{X}_2 + K_0X_2\left(1-\dfrac{L}{\sqrt{X_2^2+H^2}}\right) + K_c(X_2-X_1) + F_{s2} = 0, \end{cases} \quad (1)$$

where the dot represents the derivative with respect to time t, H is the distance between the fixed point and the conveyor belt, L is the original length of all springs, and the position of the mass i over the conveyor belt is displayed by X_i, $i = 1, 2$. The friction F_{si} between the masses and the belt is represented with the Coulomb friction. In the slipping mode, The friction F_{si} can be expressed as follows

$$\begin{cases} F_{s1} = -\mu\left[M_1g - K_0H\left(1-\dfrac{L}{\sqrt{X_1^2+H^2}}\right)\right]\mathrm{sign}(\dot{X}_1 - V_0), \\ F_{s2} = -\mu\left[M_2g - K_0H\left(1-\dfrac{L}{\sqrt{X_2^2+H^2}}\right)\right]\mathrm{sign}(\dot{X}_2 - V_0), \end{cases} \quad (2)$$

where μ represents the coefficient of friction in contacting surfaces, $F_{Ni} = M_ig - KH(1-L/\sqrt{X_i^2+H^2})$ the total contact force due to the gravity of the mass i. In this model, for the vertical component of the spring force, it is assumed that the direction of F_{Ni} always points down ($F_{Ni} > 0$) only if $M_i g > KH$. $\dot{X}_i - V_0$ denotes the relative speed of mass i and belt, $i = 1, 2$.

In the sticking mode, the speed of masses is equivalent to the speed of conveyor belt, i.e. ($\dot{X}_i = V_0$), and the acceleration of mass i is identically zero, i.e. ($\ddot{X}_i = 0$). Thus, the frictional force are obtained as

$$\begin{cases} F_{s1} = -K_0 X_1(t)\left(1 - \dfrac{L}{\sqrt{X_1^2 + H^2}}\right), \\ F_{s2} = -K_0 X_2(t)\left(1 - \dfrac{L}{\sqrt{X_2^2 + H^2}}\right). \end{cases} \tag{3}$$

Now it is supposed that the system (1) has been disturbed by an external harmonic excitation of frequency Ω and amplitude F. It is easy to get the equation of motion

$$\begin{cases} M_1 \ddot{X}_1 + K_0 X_1\left(1 - \dfrac{L}{\sqrt{X_1^2 + H^2}}\right) + K_c(X_1 - X_2) + F_{s1} = F\ \mathrm{Cos}\Omega t, \\ M_2 \ddot{X}_2 + K_0 X_2\left(1 - \dfrac{L}{\sqrt{X_2^2 + H^2}}\right) + K_c(X_2 - X_1) + F_{s2} = 0. \end{cases} \tag{4}$$

The parameters of this system become the dimensionless variables which can be expressed as follows:

$$x_i = \frac{X_i}{L},\ \alpha = \frac{H}{L},\ \frac{K_0}{M} = \omega_1^2,\ \frac{K_c}{M} = \omega_2^2,\ g_1 = \frac{g}{L\omega_1^2},\ v_0 = \frac{V_0}{L\omega_1},\ \beta = \frac{\omega_2^2}{\omega_1^2}, f = \frac{F}{KL},\ \omega = \frac{\Omega}{\omega_1},\ \tau = \omega_1 t. \tag{5}$$

Substitute Equation (5) into Equation (4), we can get the dimensionless motion equation of this system as the following :

$$\begin{cases} \ddot{x}_1 + x_1\left(1 - \dfrac{1}{\sqrt{x_1^2 + \alpha^2}}\right) + \beta(x_1 - x_2) + \mu\left[g_1 - \alpha\left(1 - \dfrac{1}{\sqrt{x_1^2 + \alpha^2}}\right)\right]\mathrm{Sign}(\dot{x}_1 - v_0) = f\ \cos\omega\tau, \\ \ddot{x}_2 + x_2\left(1 - \dfrac{1}{\sqrt{x_2^2 + \alpha^2}}\right) + \beta(x_2 - x_1) + \mu\left[g_1 - \alpha\left(1 - \dfrac{1}{\sqrt{x_2^2 + \alpha^2}}\right)\right]\mathrm{Sign}(\dot{x}_2 - v_0) = 0, \end{cases} \tag{6}$$

where the dot represents the derivative with respect to the time τ, Sign (η)= η / $|\eta|$ for $\eta \neq 0$, and Sign (0) is regarded as the whole interval [-1, 1].

3. Nonsmooth dynamical systems

3.1. Phase space topology of Filippov system

As a matter of convenience, we briefly introduce the Filippov system and bifurcation of sliding. More details are given in [11]. Assuming that the system's governing equations can be expressed as

$$\dot{x} = \begin{cases} F_1(x), & \text{if } H(x) > 0, \\ F_2(x), & \text{if } H(x) < 0, \end{cases} \tag{7}$$

where $x \in R^n$, $F_1, F_2 : R^n \to R^n$ are defined in a region D and $H : R^n \to R$ is an adequately smooth scalar function of system states. The region D is divided into two subspaces, G_1 and G_2, with continuous and smooth dynamical space. The discontinuous boundary between G_1 and G_2 is assumed to be a smooth and hyperplane Σ. Define

$$\begin{aligned} &G_1 := \{x \in R^n : H(x) > 0\}, \\ &G_2 := \{x \in R^n : H(x) < 0\}, \\ &\Sigma := \{x \in R^n : H(x) = 0\}. \end{aligned} \tag{8}$$

The switching manifold $\hat{\Sigma} \subseteq \Sigma$, which is attractive from both sides simultaneously in regions G_1 and G_2 is named as the sliding region. According to method of Utkin's equivalent control [12], we define the vector field F_S governing the flow inside the sliding region as a vector-valued function by

$$F_S = \frac{F_1 + F_2}{2} + \frac{F_2 - F_1}{2}\gamma(x), \tag{9}$$

where $\gamma(x) \in [-1, 1]$. Since F_S is tangent to the switching manifold, i.e., $\langle H_x, F_S \rangle = 0$, so

$$H_x \cdot \frac{F_1 + F_2}{2} + H_x \cdot \frac{F_2 - F_1}{2}\gamma(x) = 0. \tag{10}$$

$\gamma(x)$ is expressed in terms of F_1 and F_2 from Equation (10) as follows

$$\gamma(x) = -\frac{H_x \cdot (F_2 + F_1)}{H_x \cdot (F_2 - F_1)}. \tag{11}$$

We denote the sliding region by

$$\hat{\Sigma} := \{x \in \Sigma \mid -1 \leq \gamma(x) \leq 1\}. \tag{12}$$

3.2. Sliding bifurcations

According to the theory given in [11], sliding bifurcation is considered as tangential interactions between the edge of the sliding region and a limit cycle in Filippov systems. Due to the interaction scenarios, four possible cases can be identified and given in Figure 3, where the shadow area in the picture denotes the sliding region. Different scenarios displayed in Figure 3(a)-(d) are named as bifurcations of grazing-sliding, crossing-sliding, switching-sliding and adding-sliding, respectively. The description of these bifurcation scenarios in more detail can be given in [11].

With a parameter changing, a sliding part of the phase path reaches the tangential edges with the additional variation of the parameter, and an additional segment of the phase path lies (or below) the sliding region.

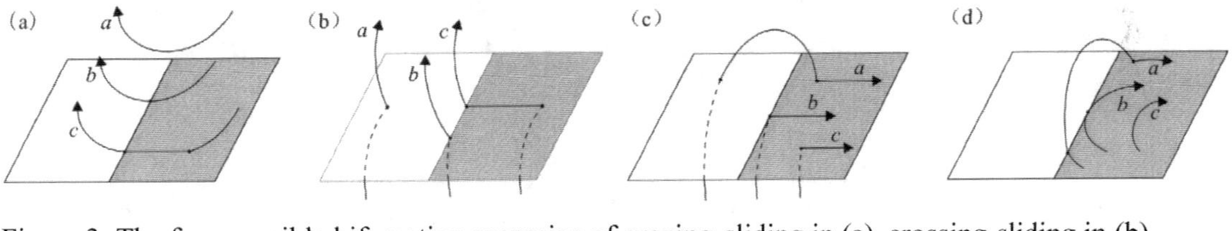

Figure 3. The four possible bifurcation scenarios of grazing-sliding in (a), crossing-sliding in (b), switching-sliding in (c) and adding-sliding in (d).

Now we consider the two DOF self-excited SD oscillators (6), and it is also a Filippov system obviously and can be expressed as follows

$$\dot{x} = \begin{cases} F_1(x), & \text{if } H_1(x) > 0 \text{ and } H_2(x) > 0, \\ F_2(x), & \text{if } H_1(x) < 0 \text{ and } H_2(x) > 0, \\ F_3(x), & \text{if } H_1(x) > 0 \text{ and } H_2(x) < 0, \\ F_4(x), & \text{if } H_1(x) < 0 \text{ and } H_2(x) < 0, \end{cases} \tag{13}$$

where $x = (x_1, y_1, x_2, y_2, \tau)^T \in R^5$, $H_1(x) = y_1 - v_0$, $H_2(x) = y_2 - v_0$ and

$$F_1(x)=\begin{bmatrix} y_1 \\ f\cos\omega\tau - x_1\left(1-\frac{1}{\sqrt{x_1^2+\alpha^2}}\right)-\beta(x_1-x_2)-\mu\left[g_1-\alpha\left(1-\frac{1}{\sqrt{x_1^2+\alpha^2}}\right)\right] \\ y_2 \\ -x_2\left(1-\frac{1}{\sqrt{x_2^2+\alpha^2}}\right)-\beta(x_2-x_1)-\mu\left[g_1-\alpha\left(1-\frac{1}{\sqrt{x_2^2+\alpha^2}}\right)\right] \\ 1 \end{bmatrix},\ F_2(x)=\begin{bmatrix} y_1 \\ f\cos\omega\tau - x_1\left(1-\frac{1}{\sqrt{x_1^2+\alpha^2}}\right)-\beta(x_1-x_2)+\mu\left[g_1-\alpha\left(1-\frac{1}{\sqrt{x_1^2+\alpha^2}}\right)\right] \\ y_2 \\ -x_2\left(1-\frac{1}{\sqrt{x_2^2+\alpha^2}}\right)-\beta(x_2-x_1)-\mu\left[g_1-\alpha\left(1-\frac{1}{\sqrt{x_2^2+\alpha^2}}\right)\right] \\ 1 \end{bmatrix},$$

$$F_3(x)=\begin{bmatrix} y_1 \\ f\cos\omega\tau - x_1\left(1-\frac{1}{\sqrt{x_1^2+\alpha^2}}\right)-\beta(x_1-x_2)-\mu\left[g_1-\alpha\left(1-\frac{1}{\sqrt{x_1^2+\alpha^2}}\right)\right] \\ y_2 \\ -x_2\left(1-\frac{1}{\sqrt{x_2^2+\alpha^2}}\right)-\beta(x_2-x_1)+\mu\left[g_1-\alpha\left(1-\frac{1}{\sqrt{x_2^2+\alpha^2}}\right)\right] \\ 1 \end{bmatrix},\ F_4(x)=\begin{bmatrix} y_1 \\ f\cos\omega\tau - x_1\left(1-\frac{1}{\sqrt{x_1^2+\alpha^2}}\right)-\beta(x_1-x_2)+\mu\left[g_1-\alpha\left(1-\frac{1}{\sqrt{x_1^2+\alpha^2}}\right)\right] \\ y_2 \\ -x_2\left(1-\frac{1}{\sqrt{x_2^2+\alpha^2}}\right)-\beta(x_2-x_1)+\mu\left[g_1-\alpha\left(1-\frac{1}{\sqrt{x_2^2+\alpha^2}}\right)\right] \\ 1 \end{bmatrix}.$$

In Filippov's representation, the self-excited system (13) has two switching surfaces $\Sigma_1 : H_1(x) = y_1 - v_0 = 0$ and $\Sigma_2 : H_2(x) = y_2 - v_0 = 0$. The switching surface representation is shown in Figure 4, with F_1 to F_4 as the smooth functions.

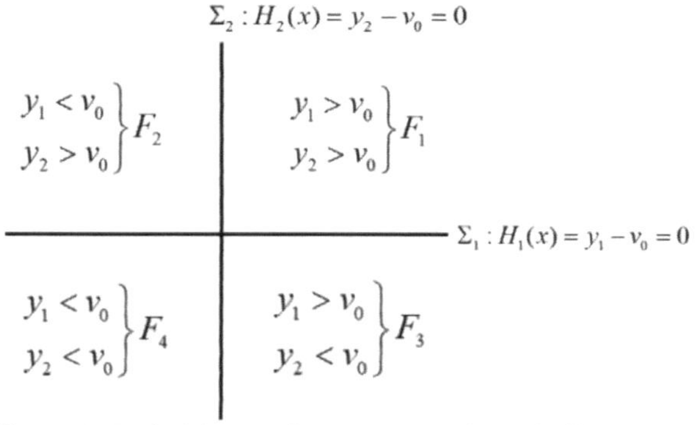

Figure 4. Switching surface presentation of Filippov system.

Since $H_{1x}(x) = (0, 1, 0, 0, 0)^T$ and $H_{2x}(x) = (0, 0, 0, 1, 0)^T$, and from the Equation (11), we can have

$$\begin{cases} \gamma_1(x) = -\dfrac{H_{1x}\cdot(F_2+F_1)}{H_{1x}\cdot(F_2-F_1)} = \dfrac{f\cos\omega\tau - x_1\left(1-\frac{1}{\sqrt{x_1^2+\alpha^2}}\right)-\beta(x_1-x_2)}{\mu\left[g_1-\alpha\left(1-\frac{1}{\sqrt{x_1^2+\alpha^2}}\right)\right]}, \\ \gamma_2(x) = -\dfrac{H_{2x}\cdot(F_4+F_3)}{H_{2x}\cdot(F_4-F_3)} = \dfrac{-x_2\left(1-\frac{1}{\sqrt{x_2^2+\alpha^2}}\right)-\beta(x_2-x_1)}{\mu\left[g_1-\alpha\left(1-\frac{1}{\sqrt{x_2^2+\alpha^2}}\right)\right]}. \end{cases} \tag{14}$$

Sliding regions of the system can be given as

$$\begin{cases} \hat{\Sigma}_1 := \{x\in\Sigma_1 \mid -1\le\gamma_1(x)\le 1\}, \\ \hat{\Sigma}_2 := \{x\in\Sigma_2 \mid -1\le\gamma_2(x)\le 1\}. \end{cases} \tag{15}$$

4. Dynamical behaviors in the system

In this section, a numerical investigation of the self-excited system leads to the detection of various dynamical behaviors. Some numerical results of the system are given to illustrate the different scenarios of sliding bifurcation. Herein, we focus on the effect of the external harmonic excitation of amplitude f and let other parameters fixed, $\mu = 0.05$, $\omega = 1.08$, $\beta = 1.2$, $g_1 = 2$, $\alpha = 0.6$ and $v_0 = 0.15$. By applying the Runge-kutta integration method on the Mathematica software, the bifurcation diagrams of the system for the displacements x_1 and x_2 vs. the excitation force f are established as f varies from 0 to 1 in an initial condition $(x_{10}, \dot{x}_{10}, x_{20}, \dot{x}_{20}) = (0.1,\ 0,\ 0.1,\ 0)$, as shown in Figures 5 (a) and (b), which clearly show that the system has both periodic and chaotic domains where the dynamics are different.

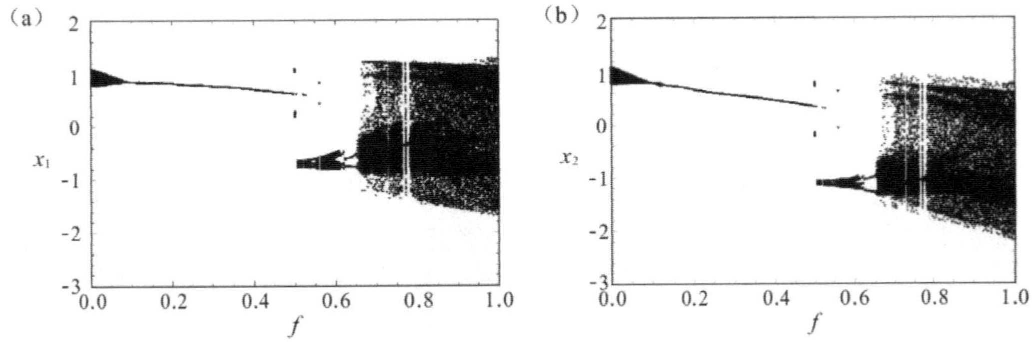

Figure 5. Bifurcation diagrams of the system in (a) for x_1 and in (b) for x_2 both versus f.

Figure 6 and Figure 7 show the corresponding phase portraits of periodic motion for the mass x_1 and x_2, respectively, and the horizontal dashed lines denote the switching manifold $v_0 = 0.15$.

Figure 6 (a) shows the pure slip periodic motion of the mass x_1 for $f = 0.17$, where the trajectory is always below the switching manifold $v_0 = 0.15$. As the amplitude f of excitation force increases, the trajectory of periodic motion is tangent to the dashed line $\dot{x} = v_0$ for $f = 0.1991$, and a grazing-sliding bifurcation of system occurs, as shown in Figure 6 (b). As f increases further, the stick-slip periodic motion occurs, and the trajectory of slipping is always below the sticking for $f = 0.31$, as presented in Figure 6 (c). For larger values of f, a new slip trajectory appears in the periodic solution with a speed greater than the speed of belt, as shown in Figure 6 (d) for $f = 0.36$, and the system undergoes a switching-sliding bifurcation. At this moment, the periodic stick-slip motion consists of a stick trajectory which terminates at the right end of the sticking interval, and a slip trajectory with a negative relative speed $(\dot{x} < v_0)$ is generated by a stick-slip transition. The slip trajectory ends up at a slip-slip transition and a new slip trajectory is started with positive relative speed $(\dot{x} > v_0)$, which ends up at a slip-stick transition in the sticking interval. In this case, the trajectory of periodic motion consists of a sequence of stick phase, plus slip phases $(\dot{x} < v_0)$ and other slip phases $(\dot{x} > v_0)$. As f increases to $f = 0.4472$ in Figure 6 (e), the transition of slip-stick motion shifts to the right and overlaps with the right end of the sticking interval in the phase plane, where a bifurcation of crossing-sliding occurs. As f further increases, only a pure slip periodic motion occurs in Figure 6 (f) for $f = 0.46$, and the trajectory is a sequence of slip phases in which $\dot{x} > v_0$ and slip phases in which $\dot{x} < v_0$.

Similarly, as we can see from Figures 7 (a)-(f), with the increasing of the excitation amplitude f, the system performs an identical dynamical behavior with sliding bifurcation for mass x_2, and bifurcations of grazing sliding, switching sliding and crossing sliding occur successively.

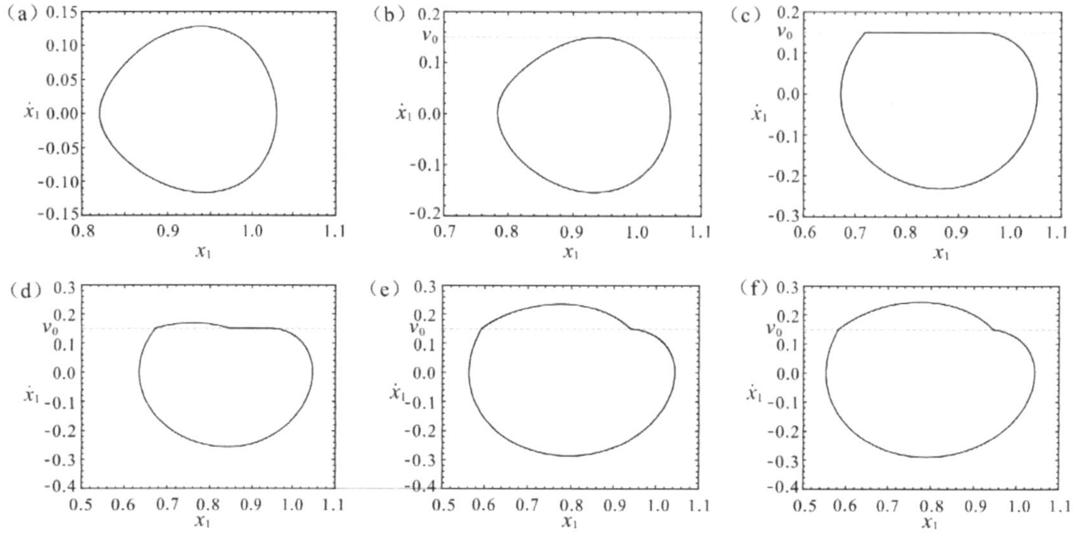

Figure 6. Sliding bifurcations of the mass 1.

(a) $f = 0.17$; (b) $f = 0.1991$; (c) $f = 0.31$; (d) $f = 0.36$; (e) $f = 0.4472$; (f) $f = 0.46$.

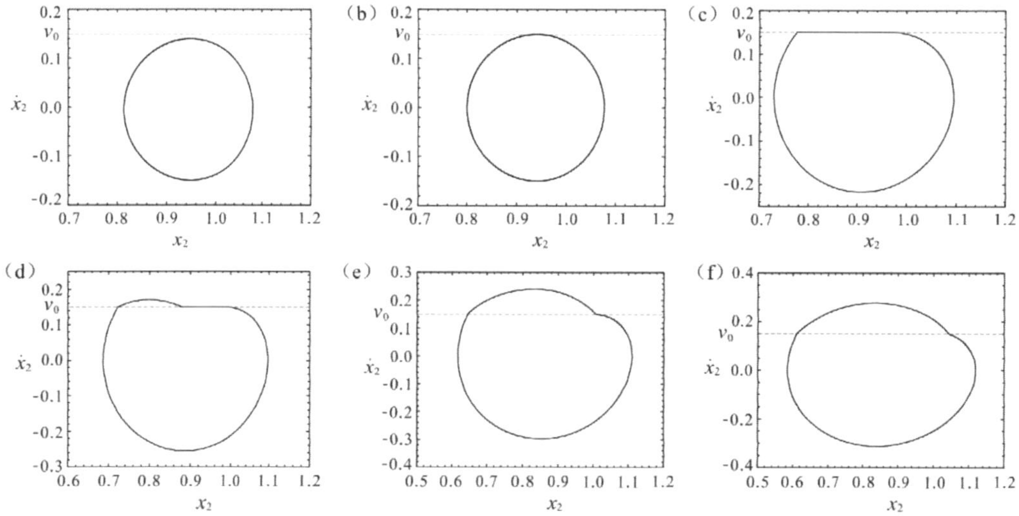

Figure 7. Sliding bifurcations of the mass 2.

(a) $f = 0.092$; (b) $f = 0.0943$; (c) $f = 0.14$; (d) $f = 0.17$; (e) $f = 0.2106$; (f) $f = 0.24$.

As the excitation amplitude f increases, the regular motion disappears, meanwhile the multiple stick-slip chaotic motions follow for $f = 0.9$, and the corresponding trajectory of chaotic motion and chaotic attractor are depicted in Figure 8 for mass 1 and in Figure 9 for mass 2.

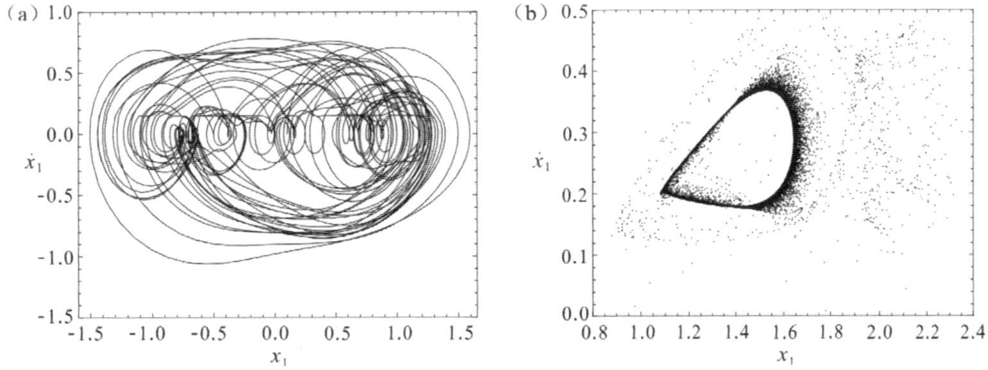

Figure 8. Chaotic motion of the mass 1 for $f = 0.9$. (a) The trajectory of chaotic motion; (b) chaotic attractor.

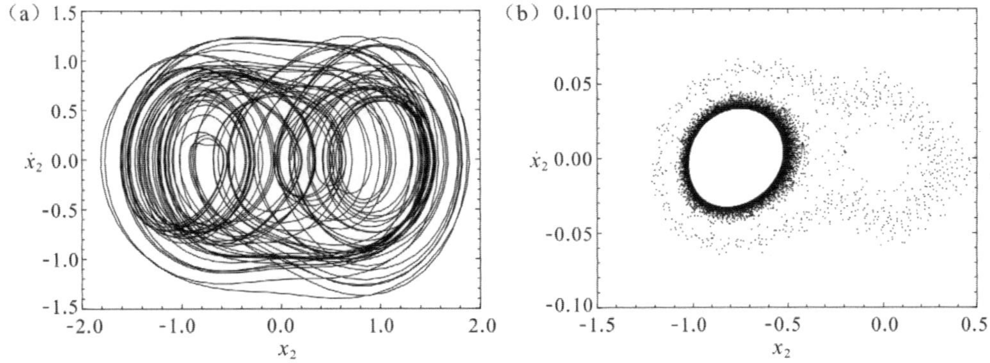

Figure 9. Chaotic motion of the mass 2 for $f = 0.9$. (a) The trajectory of chaotic motion; (b) chaotic attractor.

5. Conclusions

We have analyzed the sliding bifurcations of a two DOF self-excited SD oscillator driven by the friction of conveyor belt. The characteristic of friction is represented with the Coulomb friction. By employing the Filippov system theory, the switching surfaces and sliding regions of this nonsmooth system have been obtained. Phase portraits and bifurcation diagrams have been presented to demonstrate the various types of sliding bifurcation. The results clearly show that the system has both periodic and chaotic domains.

Acknowledgments

The authors would like to acknowledge the financial support from the Natural Science Foundation of Hebei Province, China (Grant Nos: A2019402043, A2021202013), and the Science and Technology Project of Hebei Education Department, China (Grant No. ZD2020130).

References

[1] Wiercigroch M and Krivtsov A 2001 *Philosophical Transactions of the Royal Society A* **359** 713
[2] Ouyang H, Cao Q J, Mottershead J E and Treyde T 2003 *Proceedings of the Institution of Mechanical Engineers Part D: Journal of Automobile Engineering* **271** 867
[3] Mihajlovic N, Van de Wouw N, Hendriks M P M and Nijmeijer H 2006 *Nonlinear Dynamics* **46** 273
[4] Burridge R and Knopoff L 1967 *Bulletin of the Seismological Society of America* **57** 341
[5] Li Z X, Cao Q J and Léger A 2016 *Chinese Physics B* **25** 010502

[6] Cao Q J, Wiercigroch M, Pavlovskaia E E, Thompson J M T and Grebogi C 2006 *Physical Review E* **74** 046218
[7] Tian R L, Cao Q J and Yang S P 2010 *Nonlinear Dynamics* **59** 19
[8] Li Z X, Cao Q J and Léger A 2017 *International Journal of Bifurcation & Chaos* **27** 1750009
[9] Li Z X, Cao Q J and Léger A 2017 *Nonlinear Dynamics* **89** 91
[10] Li Z X, Cao Q J and Nie Z R 2020 *Nonlinear Dynamics* **102** 1419
[11] di Bernardo M, Kowalczyk P and Nordmark A 2002 *Physica D* **170** 175
[12] Utkin V I 1992 *Sliding modes in control optimization* (Berlin: Springer-Verlag)

Dynamic modeling and analysis of compound pendulum motion of projectile

Zhang Hang, Chen Dong, Jiang Binan and Meng Qi

PLA Army Academy of Artillery and Air Defense

1850007415@qq.com

Abstract. The influence of air lift on the stability of inaccurate precision strike ammunition is studied, and the mechanical analysis of projectile and its flight stage is carried out. The dynamic equation of projectile under the action of air lift is established and simulated. The angular velocity and angular time-history curve of the compound pendulum motion of the projectile are obtained, and the correctness of the mechanical analysis of compound pendulum motion during flight is verified.

1. Introduction

Traditional artillery mainly uses indirect aiming and direct aiming when performing combat missions. The blind target are fired at intervals under the command of observation stations, giving weapons and equipment and operators time to prepare for firing. According to the relationship between the explosion point and the target position, it usually takes at least 2-3 test shots. The gun aims at the target by direct shooting means that we have been exposed to the enemy's fire[1]. Reducing the aiming accuracy to shorten the fire preparation time can damage the enemy's armor, improve the survival probability of one's own side, and achieve the purpose of seizing the enemy's initiative and mastering the battlefield initiative under actual combat[2].

Non-precision aiming precision strike ammunition is based on the existing artillery ammunition and launch platform. Without changing the operation habits and training methods, the reaction speed is fast, the influence of wind is small, and it is not easy to counter-attack[3]. It is a real low-cost design renovation. From the point of view of cost-effectiveness ratio, it is very valuable, and it can increase the combat effectiveness several times, which is of great operational significance.

2. Working principle and platform composition

When the artillery of our army finds the enemy armored target, the operator can roughly aim at the target direction after loading, and then launch the projectile, which will find the target position and launch the warhead for damage, thus greatly shortening the shooting preparation time, reducing the psychological pressure caused by the encounter to the soldiers, and then hiding in the bunker to save their combat effectiveness after launch, as shown in Figure 1.

Fig. 1. Schematic diagram of working principle.

Non-precision aiming precision strike ammunition is based on a certain type of recoilless gun. Sensor device, EFP warhead, central controller, power supply unit and other identification and detection devices are installed in the inner cavity of the projectile, forming a stable scanning platform. In order to meet the requirements of the projectile targeting accuracy and effective range, the wind cap has a conical structure and the tail adopts a movable tail structure. The torsion spring and the centrifugal force of rotation can be used to open the tail wing, and the sliding elastic belt can ensure that the rotating speed of the projectile is compatible with the flying speed.

After the projectile is shot out of the muzzle, the tail wing opens in place under the combined action of torsion spring resistance, air pressure and centrifugal force, and the projectile makes a compound pendulum motion under the action of gravity, forming a stable scanning platform, scanning and detecting the area below the projectile, and the information processing mechanism on the projectile calculates the missile-target relationship. When the projectile flies over the area above the target, the armored target is identified, and the information processing mechanism on the projectile makes a certain rule calculation according to the existing rotational speed, the preset installation angle of the damage device and the preset installation angle of the detection device, and then detonates the EFP device at the appropriate position to destroy the target. If the target is not found, it will self-destruct after hitting the ground[4].

3. Dynamic model of projectile

3.1. Analysis of uncontrolled projectile rotation.

The angle of attack of a projectile in flight is very small. In the flight process, projectile is mainly subjected to vertical downward gravity, lift perpendicular to the projectile axis, thrust and resistance before and after the projectile axis. When the projectile flies, the wing of the projectile is influenced by random crosswind, as shown in Figure 2. The angle between wind speed and the axis of the projectile is β and V, and the windward surface of the right wing of projectile increases and the windward surface of the left wing decreases, so the lift of the right wing increases and the lift of the left wing decreases, resulting in the projectile rolling. At this time, if the projectile adopts a four-piece tail structure, the crosswind will not only cause the projectile to roll, but also cause the projectile to yaw, and yaw and rolling are coupled with each other[5].

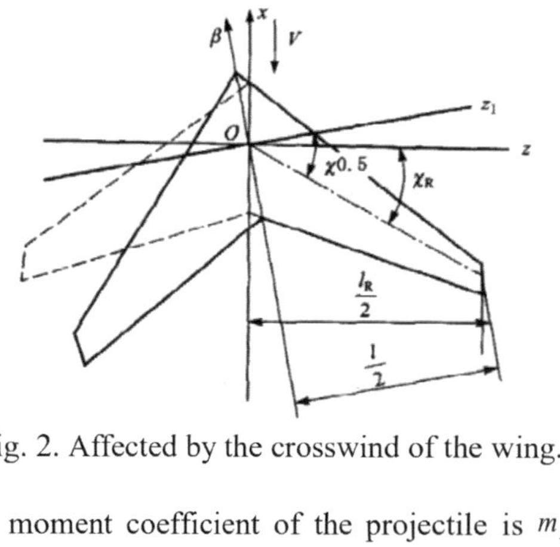

Fig. 2. Affected by the crosswind of the wing.

Assuming that the rolling moment coefficient of the projectile is m_x, the factors that affect its magnitude are: The system deviation coefficient m_{x0} caused by assembly problems, moment coefficients $m_x^{\omega_y} \bullet \omega_y$ and $m_x^{\omega_x} \bullet \omega_x$ of rolling and yawing motion caused by external factors, moment coefficient $m_x^{\beta} \bullet \beta$ caused by crosswind, and damping coefficient $m_x^{\overline{\omega_x}} \bullet \overline{\omega_x}$ caused by air resistance, namely the rolling moment coefficient m_x, can be expressed as:

$$m_x = m_{x0} + m_x^{\beta} \bullet \beta + m_x^{\omega_y} \bullet \omega_y + m_x^{\omega_x} \bullet \omega_x + m_x^{\bar{\omega}_x} \bullet \overline{\omega_x} \tag{1}$$

It can be seen from the above analysis that once the centrosymmetric projectile rolls, its spin reduction factor is only the damping moment coefficient, so it is difficult to achieve uncontrolled spin reduction for centrosymmetric projectile.

To sum up, consider the symmetric structure, or the aircraft structure, as shown in Figure 3. For projectiles with symmetrical structure, the flying direction is perpendicular to the paper surface. At this time, if the projectile has a rolling angle of γ, the left wing is up, the right wing is down, and the total lift Y exerts a lateral force on the vertical tail to the right wing, which will cause the warhead to lean to the left.

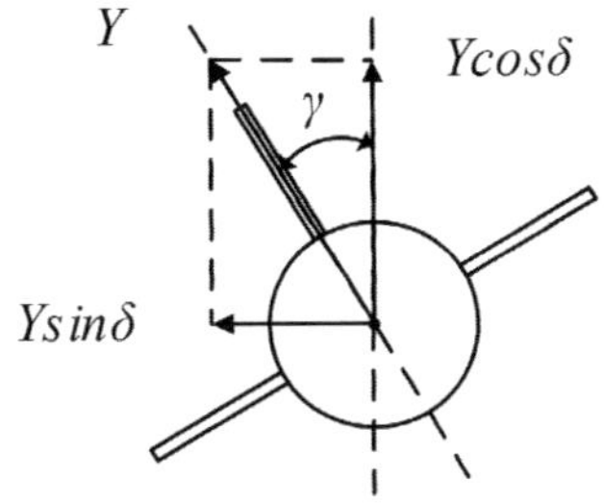

Fig. 3. Simplify the shape of projectile tail with symmetric structure.

In engineering practice, the main factors affecting the deviation moment coefficient m_{x0} of rocket rotation system are: the gap between the rocket and the artillery, the mass eccentricity of the rocket, the thrust offset, the synchronization of the inclined tail and the open tail, etc.

(1) Rotational torque M_G caused by mass eccentricity.

Let the distance between the rocket center of mass G and the rocket axis be r_G. The angle between r_G and Y axes is β. Then the tangential force caused by r_G is:

$$G_\tau = G \sin \beta \tag{2}$$

The rotating moment of the tangential force G_τ to the longitudinal axis of the projectile is:

$$M_G = G \sin \beta \cdot r_G \tag{3}$$

(2) Steering moment M_P caused by thrust offset.

Let the angle between the thrust P and the axis of projectile be α_p, and the tangential force caused by α_p is:

$$P_\tau = P\alpha_p \tag{4}$$

The rotating moment of the tangential force on the longitudinal axis of the projectile is:

$$M_P = P\alpha_p \cdot r_p \tag{5}$$

Where r_p is the eccentricity of thrust P.

(3) Steering moment M_{w1} generated by the inclined tail.

Let the angle between the tail and the missile axis be δ_2, and the normal force caused by δ is:

$$F_\delta = C_{yTW}^{\alpha} \delta_2 \cdot \frac{1}{2} \rho v^2 s_w \tag{6}$$

The rotating moment of the normal force F_δ to the longitudinal axis of the projectile is:

$$M_{w1} = F_\delta \cdot r_w \tag{7}$$

r_w is the distance between the pressure center of the tail and the elastic axis.

(4) The turning moment caused by the unsynchronization of tail.

Before and after the tail is opened, the distance between the center of mass G_w of the tail and the axis of the projectile changes, and the tail opened first will generate a turning moment on the longitudinal axis of the projectile. The distance between the center of mass G_w to the missile axis before and after opening the tail is Δr_{Gw} , and the angle between $\vec{r}_{GW}$ and Y axis is β_W . So the tangential force caused by Δr_{Gw} is:

$$G_{\tau w} = G_w \sin \beta_w \tag{8}$$

The rotating moment of the tangential force $G_{\tau w}$ to the longitudinal axis of the projectile is:

$$M_{Gw} = G_w \sin \beta_w \cdot \Delta r_{Gw} \tag{9}$$

In addition, the moment of rotation of the wing lift to the projectile's longitudinal axis is:

$$M_{w2} = C_{yTW}^{\alpha} \delta_1 \cdot \frac{1}{2} \rho v^2 s_w r_w \tag{10}$$

(5) Extremely restraining moment produced by projectile

$$M_{xzB} = \frac{1}{2} \rho v s_M l r_M m_{xzB}^{\varpi} \omega \tag{11}$$

s_M is the maximum cross-sectional area of projectile, l is the length of the projectile, and r_M is the maximum radius of the projectile.

(6) Extremely restraining moment produced by wing

Make the polar moment of inertia of the bullet C ; The muzzle velocity of the bullet is ω_0 and the angle of attack of the bullet is δ_1 ;The wing tilt angle is δ_2 , the distance between the wing pressure center to the elastic axis is r_w , and the time difference of the wing opening is Δt ;The mass eccentricity r_G , the angle formed by the mass eccentricity and the gravity line (referred to as the mass eccentricity angle for short) is β .So the rotation equation of projectile rotation is:

$$\begin{cases} C\dfrac{d\omega}{dt} = M_G + M_{Gw} + M_{w1} + M_{w2} - M_{xzB} - M_{xzw} \\ \dfrac{d\beta}{dt} = \omega \end{cases} \tag{12}$$

As shown in Figure 4, the position of the center of mass is closely related to aerodynamic shape. If the projectile moves outward and rolls in the vertical aiming direction, with the right wing on the top and the left wing on the bottom, at this time, due to the eccentric layout, the rolling moment arm of the left wing is larger than that of the right wing, so there is a balance moment to make the projectile return to the equilibrium state. Therefore, the center of mass is designed under the axis, and the specific center of mass position needs to be determined by multiple rounds of simulation and testing. Based on the principle of minimizing changes, I-shaped fixed straight wing, flat vertical tail and eccentric design scheme is adopted, and the balance tail is changed from four pieces to two pieces, and the spring leaf wing is changed to straight wing, which can overcome the adverse effect that the elastic wing can't be opened at the same time. A long vertical tail is arranged below the tail rod to realize the rolling balance.

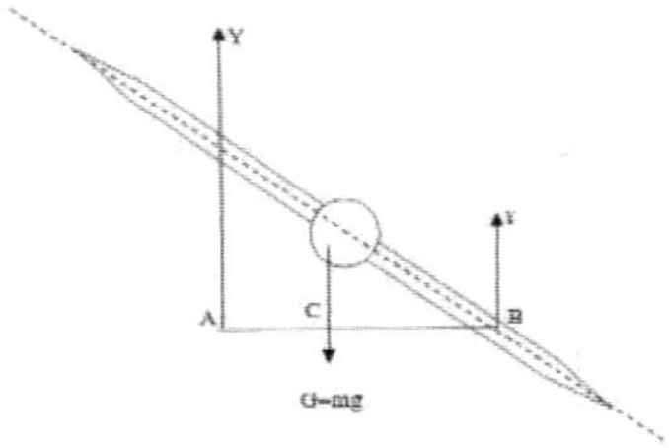

Fig. 4. Low barycenter layout.

4. Computational simulation

According to the projectile rotation equation and projectile parameters in front, the simulation calculation is carried out. The simulation conditions are: the initial angular velocity is , the initial attack angle of the bullet is , the average tilt angle of the wing is , the initial velocity is , and the center of mass axis is 3 mm. The simulation results are shown in figure 5. It can be seen that the rolling angle velocity of the projectile oscillates and balances within the effective distance, with the maximum rolling angle amplitude of 45.

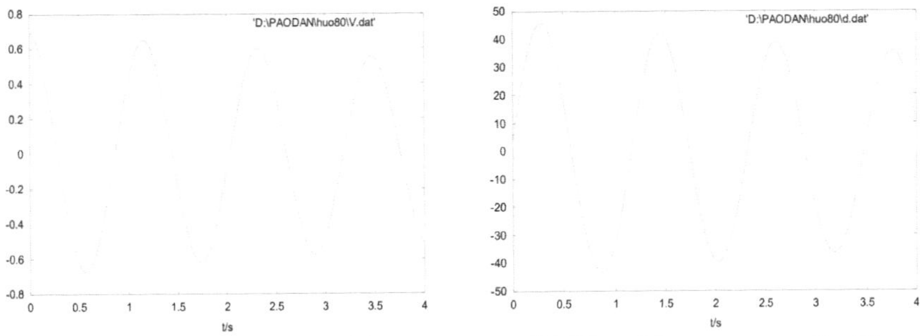

（a）Rolling angular velocity time history curve. (b) Time history curve of rolling angle.

Fig. 5. Dynamic simulation of rolling

5. Conclusion opinion

A new type of compound pendulum intelligent ammunition is introduced. By moving down the center of gravity, the projectile can rotate effectively in a short time and make a compound pendulum movement. Through the mechanical analysis of the projectile, a dynamic model is established, and the maximum angle of the projectile's compound pendulum motion is obtained through simulation calculation, which is of great significance for controlling the projectile's flight stability and obtaining accurate detection signals.

6. References

[1] DUAN Xiao-min, LI Jie, LIU Ju: Research on the Dynamic Model of a Partial Strap down Platform and the Impact Analysis of Pitching Angle and the Stability of Platform. Sep. 35(9), 1436–1442 (2014).

[2] DUAN Xiao-min, LI Jie, LIU Jun: Influence of Air Lift on the Stability of Passive Partial Strap down Platform. Nov.35(11), 1814-1819(2014)

[3] Chen Liang: Study on the Dynamic Characteristics and Motion Stability of the Hedgehopping Smart Munition. Nanjing University of Scince and Technology. jun. (2018)

[4] Chen Liang, Liu Rongzhong, Guo Rui: Aerodynamic Characteristic Study Towards the Hedgehopping Terminal Sensitive Projectile at Rolling State. Apr.38(4),338-346(2018).

[5] Zhao Bobo: Study on the Dynamic Characteristics and Hit Probability of the Hedgehopping Terminal Sensitive Projectile with High Rotational Speed. Nanjing University of Science & Technology, 2016

Consistency of Two Plastic Strengthening Parameters from EH36 Steel

Xianru LIU*, Jianguang LI, Yi REN, Kuangjie ZHAO

(Civil engineering College, Southwest Forestry University, Kunming 650000, China)

liu20785393@163.com

Abstract: In order to study the consistency of the two plastic hardening parameters applied to the constitutive model, EH36 marine steel was selected as the experimental material in this paper, and the stress-strain relationship curve of EH36 steel was obtained through uniaxial tensile experiments at room temperature, and elastic-power hardening was used. The model fits the stress-strain curve of EH36 steel and obtains the fitted parameters. Then we select a certain plastic strain arbitrarily, take the effective plastic strain and plastic work as the hardening parameters, respectively, and use the kinematic hardening model to calculate the one-dimensional stress-strain relationship of the reverse loading of the two hardening parameters when the pre-strain is applied. The consistency of the two plastic hardening parameters is demonstrated according to the calculated relationship.

1. Introduction

The manufacturing industry is an important component of a country's industrial system and one of the important signs to measure a country's scientific and technological level and comprehensive national strength [1]. In recent years, due to the continuous influence of economic globalization, the industry has developed, which has also driven the prosperity of the international shipping market. With the development of various mineral resources in the ocean, the application field of ships has been further expanded [2]. Metal is one of the main materials used in manufacturing. As we all know, Stainless steel has good corrosion resistance and low cost, which is one of the most common metals in life [3-4]. As for ships, it is required to bear the impact of wind and waves and alternate loads during sailing, so it is very strict about using steel for hull structure. EH36 steel, as one of the main shipbuilding materials, has the advantages of high strength, high toughness, and good processability.

In this paper, EH36 marine steel is used to make the specimen. With reference to the room temperature experiment method, the specimen is subjected to a uniaxial tensile test, and its stress-strain relationship, i.e., constitutive relation, is analyzed according to the obtained test data. The establishment of constitutive relations is not only the foundation of plastic mechanics but also one of the main research contents of plastic mechanics [5]. The establishment of a constitutive equation generally needs theoretical analysis. It is realized by means of experiments and numerical calculations [6]. The constitutive relation of metal materials reflects the law of macro stress changing with strain under certain deformation conditions [7]. Then, the elastic-power hardening model is used to fit the stress-strain relation of uniaxial tension. Assuming the follow-up hardening model, the effective plastic strain and plastic work are taken as hardening parameters, and the one-dimensional stress-strain relation of the two hardening parameters under a certain pre-stress is calculated. It provides a reference for the study of the mechanical properties of this material.

2. Establishment of constitutive equation

When a metal is deformed, the change of stress with strain is the constitutive relation of the metal material, and the constitutive relation is the basis of studying the material deformation [8]. In order to fully express the elastic-plastic reaction, the concepts of loading rule, flow rule, strengthening rule, strengthening parameters, and compatibility conditions are very important, which are the basis of the plastic theoretical framework [9]. In a certain plastic deformation state, a load increment is applied to the object, and the stress increment and strain increment generated by the load increment is $d\sigma$ and $d\varepsilon$, respectively[10-13]. The strain increment can be divided into elastic strain increment and plastic strain increment, namely:

$$d\varepsilon = d\varepsilon^{e} + d\varepsilon^{p} \tag{1}$$

The expressions of stress increment and strain increment are:

$$d\sigma = E_t d\varepsilon \tag{2}$$

where E_t is the tangent modulus, and the expression is:

$$E_t = E\frac{-\frac{\partial f}{\partial \kappa}h}{\left(\frac{\partial f}{\partial \sigma}\right)^2 E - \frac{\partial f}{\partial \kappa}h} \tag{3}$$

The history of stress is expressed by the current value of the scalar reinforcement parameter, which is marked as κ . Its increment expression is:

$$d\kappa = h d\lambda \tag{4}$$

where h is a scalar function.

For effective plastic strainε_p:

$$h = \sqrt{\frac{\partial f}{\partial \sigma}\frac{\partial f}{\partial \sigma}} \tag{5}$$

For plastic work W_P:

$$h = \sigma\frac{\partial f}{\partial \sigma} \tag{6}$$

In this paper, the following reinforcement law is adopted, and the loading function related to the following reinforcement law is mathematically expressed as:

$$f = (\sigma, \alpha) = (\sigma - \alpha)^2 - \sigma_s^2 \tag{7}$$

where α is the reaction force, and σ_s is the initial yield stress. The reaction force α is also a function of the plastic loading history, and the incremental form is expressed as:

$$d\alpha = \frac{d\sigma_e}{d\kappa}\frac{d\varepsilon^{p}}{|d\varepsilon_p|}d\kappa \tag{8}$$

where σ_e is the effective stress, which reflects the way of stress changes in the loading stage[10].

3. Experiment and data fitting

3.1 Specimen design

The material selected in this paper is EH36 marine steel purchased from Wuxi Biaolong Metal Products Co., Ltd., and the base material size is 500mm×400mm×20mm.

The design of the specimen follows the standard of the Tensile Test Method for Metallic Materials, and then the base steel plate is cut into test specimens in Kunming, Yunnan Province. The size of the designed specimen is shown in Figure 1, and the physical drawing of the specimen is shown in Figure 2.

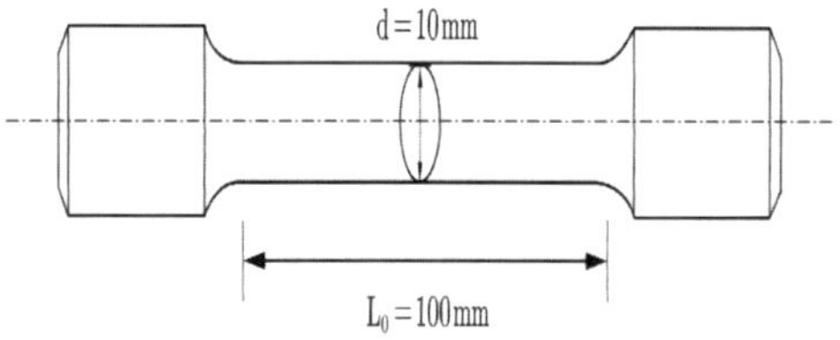

Fig. 1 Specimen design

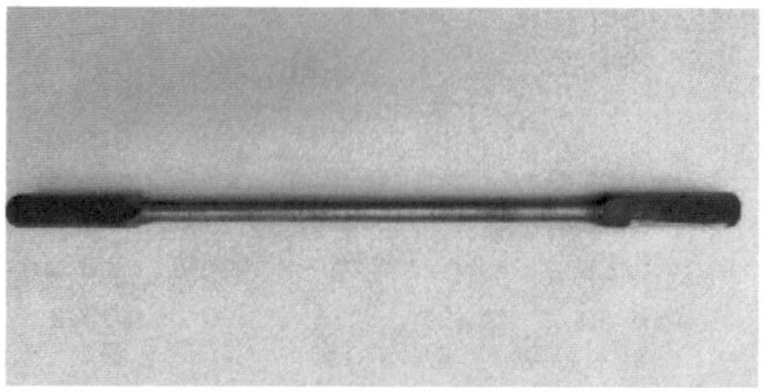
Fig. 2 Specimen physical drawing

3.2 Experiment equipment and process

Referring to the Standard GB/T228-2010 for Tensile Test Method of Metallic Materials at Room Temperature, the quasi-static tensile test conditions of EH36 steel at room temperature are set as follows: room temperature 20℃ and strain rate 2×10-4/s, and the test results are averaged, and the stress-strain curve of quasi-static tensile material at room temperature is obtained [14-15].

This test was completed in the laboratory of Wuxi Commercial Testing Technology Co., Ltd., The test equipment used is an electronic universal testing machine, and its number is BTT-INS-003. This equipment is calibrated every year, which greatly reduces the error caused by the equipment. The equipment is shown in Figure 3. The purpose of the metal uniaxial tensile test is to obtain the quasi-static stress-strain relationship of EH36 steel at room temperature. After cutting and grinding, the test pieces need to be dried and stored. During the test, the specimen needs to stand at room temperature for a period of time before the tensile test. The specimen after the test is shown in Figure 4.

Fig. 3 Experimental equipment diagram

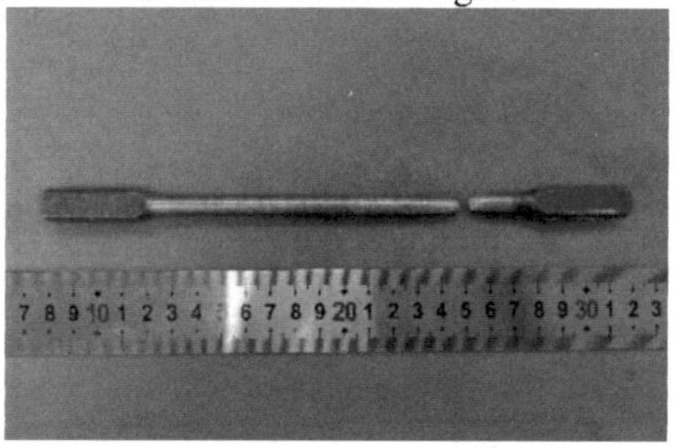
Fig. 4 Specimen diagram after the end of the test

3.3 Experimental data and analysis

Set the strain loading rate to 2.00mm/min. Use the universal material testing machine to do the tensile test on metal samples, and get the stress-strain data of materials. After finishing the test data, the stress-strain curve of EH36 steel is shown in Figure 5.

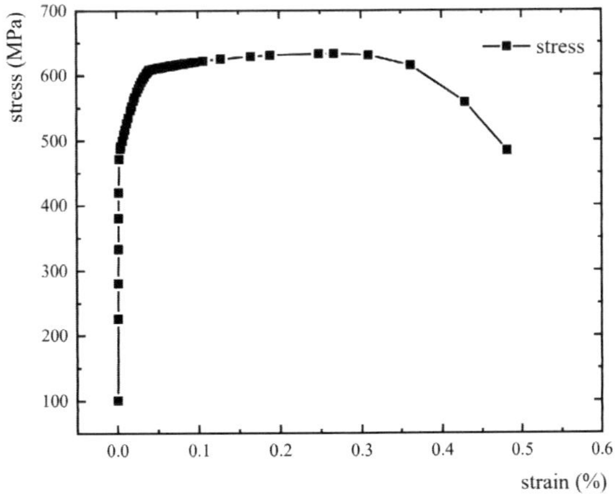

Fig. 5 Stress-strain curve

According to the experimental data, the yield strength of this material is 488 MPa, and the residual deformation stress of 0.2% is taken as the yield strength of this material. The mechanical properties of EH36 steel are shown in Table 1 below.

Table 1 Mechanical properties of EH36 steel

Yield strength /MPa	Tensile strength /MPa	Elongation after fracture /%	Section shrinkage /%	Modulus of elasticity /GPa
488	633	20.8	74	221

In order to study the mechanical properties of this material, the elastic-power hardening model is used to describe the consistency of the two hardening parameters, and its expression is:

$$\begin{aligned} \sigma &= E\varepsilon \qquad ,\sigma \leq \sigma_s \\ \sigma &= \sigma_s + m(\varepsilon^p)^n, \sigma > \sigma_s \end{aligned} \tag{9}$$

where σ_s is the stress value corresponding to the yield point of the material, m and n are the material parameters and ε^p is the plastic strain. In this paper, the experimental data are fitted to obtain parameters m and n, and the stress-strain relationship in the plastic region of the material is transformed into the stress-plastic strain relationship, as shown in Figure 6 below.

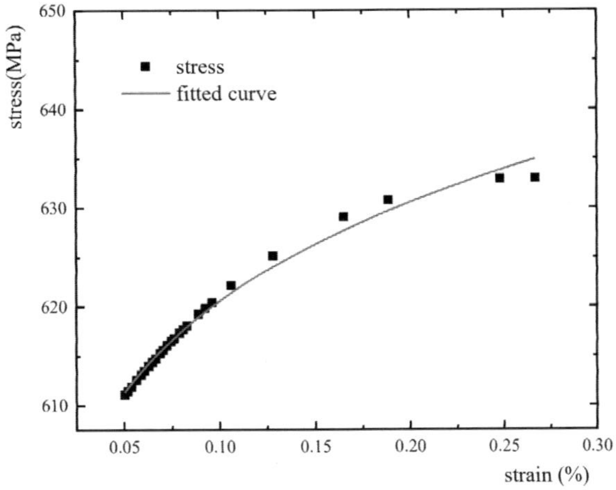

Fig. 6 Stress and plastic strain relationship diagram

According to the material parameters obtained by fitting, $m=168$, $n=0.1$, the relationship between stress and plastic strain is as follows:

$$\sigma = 488 + 168(\varepsilon^{p})^{0.1} \tag{10}$$

4. Demonstrate the consistency of the two strengthening parameters.

The plastic history is expressed by the current value of the scalar strengthening parameter. The two strengthening parameters used in this paper are effective plastic strain ε_p and plastic work, respectively W_p.

In order to demonstrate the consistency of the two strengthening parameters, according to the parameters obtained by fitting, it is assumed that the follow-up strengthening rule is adopted and the two strengthening parameters are calculated respectively. Pre-strain to $\varepsilon^{p} = 0.002$, and obtain the one-dimensional stress-strain relationship of reverse loading at this time.

For the effective plastic strain ε_p as the strengthening parameter, the stress-strain relationship of the material is as follows:

$$\varepsilon = \frac{\sigma}{221000} - 0.004 - \left(-\frac{\sigma + 307.6}{168}\right)^{1/0.1} \tag{11}$$

For plastic work W_p as a strengthening parameter, the one-dimensional stress-strain relationship can be obtained as follows:

$$\varepsilon = \frac{\sigma}{221000} - 0.004 - \left(-\frac{\sigma + 307.6}{168}\right)^{1/0.1} \tag{12}$$

5. Conclusion

It can be seen from Equation (11)and Equation (12)that when effective plastic strain and plastic work are selected as strengthening parameters, the calculated reverse stress-strain relationship is the same. According to the calculated stress-strain relationship, the consistency of the two strengthening parameters is verified.

References

[1]LI X X. Research on the Constitutive Model Parameters Identification of 304 Stainless Steel [D].Huazhong University of Science and Technology, 2012. (in Chinese)
[2]Xin C. Establishment and application of elastoplastic constitutive model for cyclic loading[D]. Yanshan University. (in Chinese)
[3] DUAN W F, DENG Z P, LIU W Y, et al. Experimental study of the constitutive model of stainless steel S30408[J]. Steel Structure, 2016(5):5. (in Chinese)
[4] WANG Y Q, CHANG Ting, SHI Y Y, et al. Experimental study of the stress-strain relation of Austenitic Stainless Steel S31608[J]. Journal of Tsinghua University: Science Natural Edition, 2013(9):5. (in Chinese)
[5] LIANG L F, LIU S Q. Elastoplastic constitutive relationship of general loading law[J]. Chinese Journal of Solid Mechanics, 2001, 22(4):6. (in Chinese)
[6] XING Y H. Research on several problems in plastic constitutive theory[D]. Harbin Engineering University, 2007. (in Chinese)
[7] PENG H B, ZHANG H J. Research development of the Constitutive Models of Metal Materials[J]. Materials for Mechanical Engineering, 2012, 36(3):7. (in Chinese)
[8] QIN Fei, WU Bin. Theoretical basis of elasticity and plasticity[M]. Science Press, 2011. (in Chinese)
[9] Chen H F, A.F. SARIPU, SARIPU, et al. Elastic and Plastic Mechanics[M]. China Architecture and Building Press, 2005. (in Chinese)
[10] CHEN M X. Elastoplastic mechanics[M]. Science Press, 2007. (in Chinese)
[11] PERZYNA, P. The constitutive equations for rate sensitive plastic materials[J]. Quart.appl.math, 1961, 20(4):321-332.
[12] ZIENKIEWICZ O C, Taylor R L, Fox D . The Finite Element Method for Solid and Structural

Mechanics. 2005.
[13] XIONG Hui. A new hardening model of metal constitutive equations[D]. Chongqing University, 2003. (in Chinese)
[14] TAO Jianhua, JIAO Ling, HE Wu. Interpretation of GB/T 228.1-2010 Tensile Test of Metallic Materials Part 1: Test Methods at Room Temperature[J]. Ship Standardization and Quality, 2013, 000(002):21-25. (in Chinese)
[15] GAO Y F, LIANG X B, DENG X L. GB/T 228.1-2010 "Tensile Testing of Metallic Materials Part 1: Experimental Methods at Room Temperature" Implementation Guidelines[M]. China Quality Inspection Press, 2012. (in Chinese)

Analytic solution on force analysis of tripod in arbitrary space

Yun LI[1*] , Yumin SONG [2]

[1.] Shandong Technology and Business University, Shandong Province, 264000, China.

[2.] School of urban rail transit, Shanghai University of engineering and technology, 201620, China

E-mail*: 58046910@qq.com

Abstract: Tripod is a common support structure. The key to tripod design is to accurately analyze the force, select appropriate materials, and design the angle according to the size of the axial force. To study the stress condition and application performance of a tripod, this paper takes any tripod in space as the research object, through the combination of mathematics and mechanics, establishes a space rectangular coordinate system, solves the internal force of the tripod member, and verifies the correctness of the solution results. The evolution from arbitrary tripod structure to special tripod structure demonstrates the correctness of the reasoning method from general to special. An arbitrary space tripod is a common engineering application structure. Taking it as the research object, mechanical analysis has certain practical significance.

1. Introduction

Tripod is a common support with a simple structure, stable support, and easy adjustment. It is widely used in weapon equipment, photographic equipment, power transmission tower, and other engineering scenes. In the actual project, with the change of the erection angle of the tripod, the stress of the support members will change. At the same time, the tripod structure is a special truss structure, which has the advantage of improving strength and stability by decomposing the external force through axial force and is favored by engineers.

Generally speaking, the special tripod structure can be solved by structural mechanics. However, it is difficult to solve the triangular frame of any form in space using structural mechanics. With the development of finite element theory and software at home and abroad, the finite element method is usually used for the stress analysis of truss structures [1~4]. Su Caihang [5] briefly introduced the typical analysis process of solving classical mechanical problems such as truss stress through the finite element software ANSYS APDL. Gao Yunfeng [6] proposed that Matlab be applied to the truss structure calculation of theoretical mechanics. Yuan Lijun [7] studied applying a stereotyped tripod single-side formwork support system in construction in combination with engineering practice. Huang Peibing [8] introduced the finite element analysis process of the force on the hoisting node by taking a steel pipe truss integral hoisting project as an example. Wen Qingjie [9] studied the mechanical performance of large eccentric joints of aluminum alloy truss bridges by combining finite element simulation and theoretical calculation methods. Yuan Anhua [10] established a full bridge spatial finite element model through Midas Civil in combination with the actual project to analyze the stress and seismic stability of the whole bridge. In a word, the finite element method is widely used in engineering practice and is an effective and reliable method.

Although the finite element method is reliable and effective, the discussion on the analytical solution of the force analysis of the tripod with arbitrary shape in space has not only certain mechanical significance but also the characteristics of mathematical application. The combination of mechanics, mathematics, and solid geometry has a good application significance of mathematics in mechanics. It can also qualitatively analyze the stress of the structure according to the analytical solution. Therefore, this paper takes any tripod in space as the research object, derives and solves the axial force of the tripod through the analysis method of combining mathematics and mechanics, and verifies the correctness of the solution content according to the calculation results under special circumstances.

2. Stress analysis of tripod support

2.1. Construction of the spatial rectangular coordinate system

Any three bars of length a, b, and c form the space truss structure, as shown in Figure 1. To obtain the internal forces of bars a, b, and c, a space rectangular coordinate system is established.

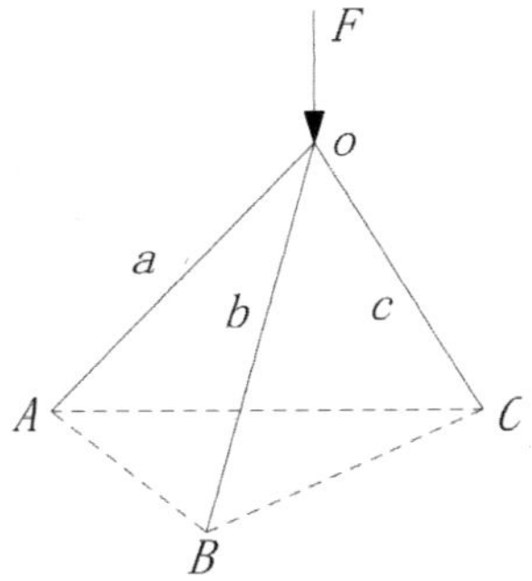

Fig. 1 Space tripod support

As shown in Figure 2, we take the vertical projection point D of point O on plane ABC as the origin to establish a space rectangular coordinate system.

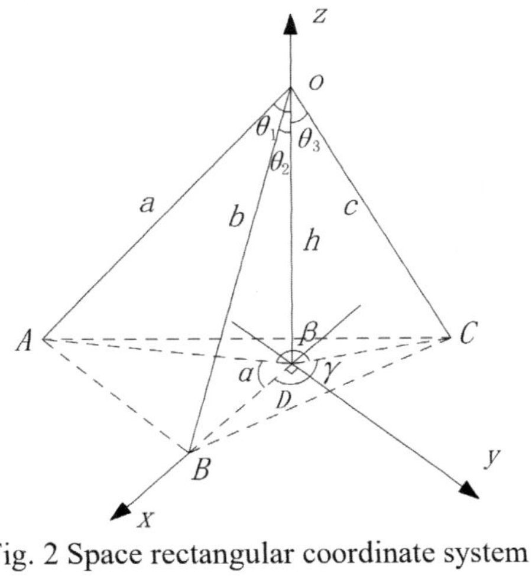

Fig. 2 Space rectangular coordinate system

2.2. Solution process

We set OD height as h, ∠ ADB= α, ∠ADC= β, ∠BDC= γ, ∠AOD= θ 1, ∠BOD= θ 2, and ∠COD= θ 3. The internal forces of rods a, b, and c are Fa, Fb, and Fc, respectively.

The following is the solution process:

$$\sum X = 0, -F_b \sin\theta_2 - F_a \sin\theta_1 \cos\alpha - F_c \sin\theta_3 \cos\gamma = 0 \tag{1}$$

$$\sum Y = 0, F_a \sin\theta_1 \sin\alpha - F_c \sin\theta_3 \sin\gamma = 0 \quad (2)$$

$$\sum Z = 0, F_a \cos\theta_1 + F_b \cos\theta_2 + F_c \cos\theta_3 = F \quad (3)$$

Above are $COS\theta_1 = \frac{h}{a}$, $\cos\theta_2 = \frac{h}{b}$, $\cos\theta_3 = \frac{h}{c}$, $\sin\theta_1 = \sqrt{1-\frac{h^2}{c^2}}$, $\sin\theta_2 = \sqrt{1-\frac{h^2}{b^2}}$, and $\sin\theta_3 = \sqrt{1-\frac{h^2}{c^2}}$.

The solution is:

$$Fa = \frac{F}{\cos\theta_1 - \frac{\sin\theta_1 \cos\alpha \cos\theta_2}{\sin\theta_2} - \frac{\sin\theta_1 \sin\alpha \cos\gamma \cos\theta_2}{\sin\gamma \sin\theta_2} + \frac{\cos\theta_3 \sin\theta_1 \sin\alpha}{\sin\theta_3 \sin\gamma}} \quad (4)$$

$$F_b = -\frac{F}{\frac{\sin\theta_2 \sin\gamma \cos\theta_1}{\sin\theta_1 \sin\alpha} - \frac{\sin\gamma \cos\alpha \cos\theta_2}{\sin\alpha} - \cos\gamma \cos\theta_2 + \frac{\sin\theta_2 \cos\theta_3}{\sin\theta_3}} \left(\frac{\sin\gamma \cos\alpha}{\sin\alpha} + \cos\gamma \right)$$

(5)

$$F_c = \frac{F}{\frac{\sin\theta_3 \sin\gamma \cos\theta_1}{\sin\theta_1 \sin\alpha} - \frac{\sin\theta_3 \sin\gamma \cos\alpha \cos\theta_2}{\sin\alpha \sin\theta_2} - \frac{\sin\theta_3 \cos\gamma \cos\theta_2}{\sin\theta_2} + \cos\theta_3} \quad (6)$$

The simplification is:

$$F_a = \frac{Fa}{h\left(1 - \cos\alpha\sqrt{\frac{a^2-h^2}{b^2-h^2}} - \frac{\sin\alpha \cos\gamma}{\sin\gamma}\sqrt{\frac{a^2-h^2}{b^2-h^2}} + \frac{\sin\alpha}{\sin\gamma}\sqrt{\frac{a^2-h^2}{c^2-h^2}}\right)} \quad (7)$$

$$F_b = -\frac{Fb}{h\left(\frac{\sin\gamma}{\sin\alpha}\sqrt{\frac{b^2-h^2}{a^2-h^2}} - \frac{\sin\gamma \cos\alpha}{\sin\alpha} - \cos\gamma + \sqrt{\frac{b^2-h^2}{c^2-h^2}}\right)} \left(\frac{\sin\gamma \cos\alpha}{\sin\alpha} + \cos\gamma \right) \quad (8)$$

$$F_c = \frac{Fc}{h\left(\frac{\sin\gamma}{\sin\alpha}\sqrt{\frac{c^2-h^2}{a^2-h^2}} - \frac{\sin\gamma \cos\alpha}{\sin\alpha}\sqrt{\frac{c^2-h^2}{b^2-h^2}} - \cos\gamma\sqrt{\frac{c^2-h^2}{b^2-h^2}} + 1\right)} \quad (9)$$

2.3 Verification of calculation results

（1）When a=b=c, ∠ADB=∠ADC=∠BDC, that is $\alpha=\beta=\gamma= 120^0$.

$$F_a = \frac{Fa}{h\left(1-\cos\alpha\sqrt{\frac{a^2-h^2}{b^2-h^2}} - \frac{\sin\alpha\cos\gamma}{\sin\gamma}\sqrt{\frac{a^2-h^2}{b^2-h^2}} + \frac{\sin\alpha}{\sin\gamma}\sqrt{\frac{a^2-h^2}{c^2-h^2}}\right)} = \frac{Fa}{h\left(1+\frac{1}{2}+\frac{1}{2}+1\right)} = \frac{Fa}{3h} \tag{10}$$

$$F_b = -\frac{Fb}{h\left(\frac{\sin\gamma}{\sin\alpha}\sqrt{\frac{b^2-h^2}{a^2-h^2}} - \frac{\sin\gamma\cos\alpha}{\sin\alpha} - \cos\gamma + \sqrt{\frac{b^2-h^2}{c^2-h^2}}\right)}\left(\frac{\sin\gamma\cos\alpha}{\sin\alpha}+\cos\gamma\right) = -\frac{Fb}{h\left(1+\frac{1}{2}+\frac{1}{2}+1\right)}\left(-\frac{1}{2}-\frac{1}{2}\right) = \frac{Fb}{3h} \tag{11}$$

$$F_c = \frac{Fc}{h\left(\frac{\sin\gamma}{\sin\alpha}\sqrt{\frac{c^2-h^2}{a^2-h^2}} - \frac{\sin\gamma\cos\alpha}{\sin\alpha}\sqrt{\frac{c^2-h^2}{b^2-h^2}} - \cos\gamma\sqrt{\frac{c^2-h^2}{b^2-h^2}} + 1\right)} = \frac{Fc}{h\left(1+\frac{1}{2}+\frac{1}{2}+1\right)} = \frac{Fc}{3h} \tag{12}$$

(2) When b=h, $\theta_2=0$,

$$F_a = \frac{Fa}{h\left(1-\cos\alpha\sqrt{\frac{a^2-h^2}{b^2-h^2}} - \frac{\sin\alpha\cos\gamma}{\sin\gamma}\sqrt{\frac{a^2-h^2}{b^2-h^2}} + \frac{\sin\alpha}{\sin\gamma}\sqrt{\frac{a^2-h^2}{c^2-h^2}}\right)} \tag{13}$$

$$\text{b=h}, \sqrt{\frac{a^2-h^2}{b^2-h^2}} \to \infty, F_a \to 0$$

$$F_b = -\frac{Fb}{h\left(\frac{\sin\gamma}{\sin\alpha}\sqrt{\frac{b^2-h^2}{a^2-h^2}} - \frac{\sin\gamma\cos\alpha}{\sin\alpha} - \cos\gamma + \sqrt{\frac{b^2-h^2}{c^2-h^2}}\right)}\left(\frac{\sin\gamma\cos\alpha}{\sin\alpha}+\cos\gamma\right) = -\frac{Fb}{h\left(0-\frac{\sin\gamma\cos\alpha}{\sin\alpha}-\cos\gamma+0\right)}\left(\frac{\sin\gamma\cos\alpha}{\sin\alpha}+\cos\gamma\right) = F \tag{14}$$

$$F_c = \frac{Fc}{h\left(\frac{\sin\gamma}{\sin\alpha}\sqrt{\frac{c^2-h^2}{a^2-h^2}} - \frac{\sin\gamma\cos\alpha}{\sin\alpha}\sqrt{\frac{c^2-h^2}{b^2-h^2}} - \cos\gamma\sqrt{\frac{c^2-h^2}{b^2-h^2}} + 1\right)}$$

(15)

$$b=h, \sqrt{\frac{c^2-h^2}{b^2-h^2}} \to \infty, F_c \to 0$$

3. Conclusion

(1) Arbitrary space tripod is a common engineering application structure. Taking it as the research object, combining mathematics and mechanics, it is of great significance in applying mathematics in mechanics to carry out mechanical analysis of triangular structures with arbitrary shapes in space.

(2) The space rectangular coordinate system is established, the axial forces Fa, Fb, and Fc of any tripod are solved, and the correctness of the solution is verified by calculating the solution results in special cases.

(3) The evolution from arbitrary tripod structure to special tripod structure demonstrates the correctness of the reasoning method from general to special.

References

[1] Li LK (2017), Structural Mechanics (I) [M]. Higher Education Press.
[2] Kang X, Yuan ZY, (2022) Comparative analysis of finite element software based on truss structure [J]. Anhui Architecture, 29 (02): 68-70.
[3] Shao H, (2018), Finite element analysis of truss structure based on ANSYS [J]. Sichuan Building Materials, 44 (09): 67-68.
[4] Lin AJ, (2018), Finite element analysis and optimal design of truss structures [J]. Engineering Technology Research, (09): 200-201
[5] Su CH, He JX, Deng JD, Wang BQ, Li C. (2019), Stress Analysis of Truss Structure Based on ANSYS [J]. Science and Technology Innovation, (29): 58-59.
[6] Gao YF. (2021), Matlab Series for Solving Theoretical Mechanical Problems (I) Stress Problems of Rigid Body System and Truss [J]. Mechanics and Practice, 43 (02): 256-261.
[7] Yuan LJ, Qi CY, Wang HH, Zhou HW, Zou HZ. (2021), Construction Technology of Stereotyped Tripod Single Side Formwork in External Wall of Deep Foundation Pit [J]. Engineering Quality, 39 (05): 23-27.
[8] Huang PB. (2021) Finite element analysis of steel pipe truss hoisting node stress [J]. Electromechanical Engineering Technology, 50 (09): 193-195.
[9] Wen QJ, Xu YH, Ren ZJ. (2021), Mechanical performance analysis and design optimization of large eccentric joint connection of aluminum alloy truss bridge [J]. Structural Engineer, 37 (05): 205-213.
[10] Yuan AH, (2021), Analysis on Mechanical Characteristics of Steel Truss Pedestrian Overpass [J]. Low Carbon World, 11 (12): 155-156.

Journal of Physics: Conference Series **2455** (2023) 012010 doi:10.1088/1742-6596/2455/1/012010

Theoretical Prediction and Experimental Verification of Hitting Targets of Gravity-Powered Trebuchets

Xiangshu Liu[1], Qicheng Chen[2], Ying Zhai[1], Lintang Mo[2], Maoli Yang[1]

[1]Department of Physics and Electrical Engineering, Liupanshui Normal University, Liupanshui, Guizhou, China, 553004

[2]College of Science, Beibu Gulf University, Qinzhou, Guangxi, China, 535000

lxsmax@163.com

Abstract: As one of the early weapons used by the ancient Chinese military, the gravity-powered trebuchet is listed as a mandatory program in science and technology games in several provinces. Due to its entertaining and informative features, this program has received continuous attention and discussion from scholars. Currently, most of the studies on this program are empirical, and few studies have been conducted quantitatively based on theoretical models. Therefore, this study established a theoretical model of the trebuchet projection process from the perspective of system energy conservation and gave an accurate analytical expression of hitting target points. In order to verify the reliability of the theoretical model, experimental tests were conducted. The results showed that the experimental results were relatively consistent with the theoretical predictions. These findings could provide a new research idea for facilitating gravity-powered trebuchets to hit targets accurately.

1. Introduction

As an indispensable weapon for long-range attack and defense in ancient battlefields at home and abroad, the trebuchet has had a prominent place in the history of weapons for more than 2,000 years, from its appearance to its retirement[1-5]. According to the literature, previous studies on trebuchets are mainly focused on two aspects. Scholars have mainly studied the types, structure, performance, and development history of trebuchets. In recent years, some researchers have made valuable research achievements by taking advantage of computer simulation software. However, few studies have been conducted quantitatively based on theoretical models from a physical view[6-8].

In this paper, a gravity-powered trebuchet was taken as an example to establish a theoretical model and analyze the kinetic law of the projection process from the perspective of system energy conservation and the motion law of the projectile from the perspective of kinematics. Through a series of theoretical derivations, the precise theoretical equations that are satisfied by the trebuchet in hitting the target point were obtained. In order to verify the reliability of the theoretical model, experimental tests were also conducted. The results showed that the experimental results were well consistent with the theoretical model.

2. Establishment of a Theoretical Model of Gravity-Powered Trebuchets

2.1. Construction of a gravity-powered trebuchet model

A gravity-powered trebuchet model was established (Figure 1), and it mainly consists of three parts: the base, mainframe, and working lever. The working lever, also called the throwing arm, works as a hand lever. The shorter arm of the throwing arm is the power arm, at the end of which a weight is placed to provide the throwing energy through the free fall of the counterweight, while the long arm of the throwing arm is the resistance arm, at the end of which a carrying device (usually in the shape of a flat spoon) is configured to carry the projectile. In order to achieve good throwing performance, the ratio of the length of the power arm to that of the resistance arm shall generally be within a certain range. Relevant practical experience shows that the throwing performance is optimal when the ratio is 1:5 to 1:2.5.

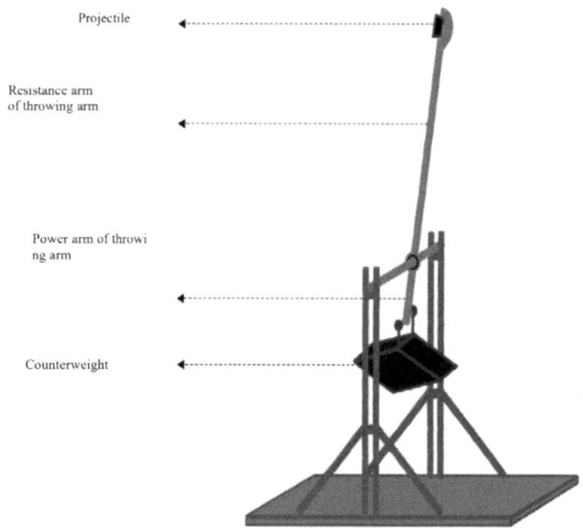

Figure 1. Gravity-powered Trebuchet

2.2. Brief analysis of the gravity-powered trebuchet model

In this section, the model was taken as the study object for theoretical analysis and in-depth study. To facilitate the following theoretical analysis, a schematic of the throwing process was made, as shown in Figure 2 below.

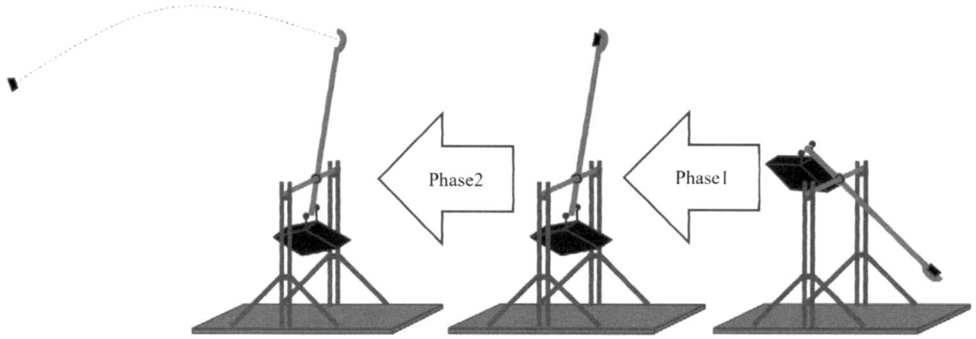

Figure 2. Schematic of the Throwing Process

The energy conversion in the throwing process involves potential energy, kinetic energy, rotational energy, frictional energy dissipation, vibration energy dissipation, etc.

3. Theoretical Analysis of Gravity-Powered Trebuchets

The theoretical analysis starts mainly from the throwing process of the gravity-powered trebuchet, which is divided into two phases. In Phase 1, the trebuchet changes from the energy storage state to the critical state, and the study object is the entire trebuchet, so this phase is called the energy phase. In Phase 2, the trebuchet changes from the critical state to the attack state, and the study object is the projectile, so this phase is called the motion phase.

3.1. Dynamics analysis of the gravity-powered trebuchet

The changes in energy of the gravity-powered trebuchet in the throwing process were obtained through analysis. The reduced energy includes the gravitational potential energy of the counterweight, the gravitational potential energy of the power arm of the throwing arm, and the vibration energy dissipation (the materials used for the construction of the trebuchet are mostly wood, bamboo, etc., which are of high elasticity.), the frictional energy dissipation of the mechanical device (it mainly refers to the energy dissipation of rolling bearing fiction), and the energy dissipation of the air friction (since air friction loss is very small, the calculation process is cumbersome, and many uncertainties exist, so this energy dissipation is ignored).

The increased energy includes the gravitational potential energy of the projectile, the gravitational potential energy of the resistance arm of the throwing arm, the kinetic energy of the projectile and the counterweight, and the rotation energy of the lever. According to the law of energy conservation and transformation, the energy transformation diagram was obtained, as shown in Figure 3.

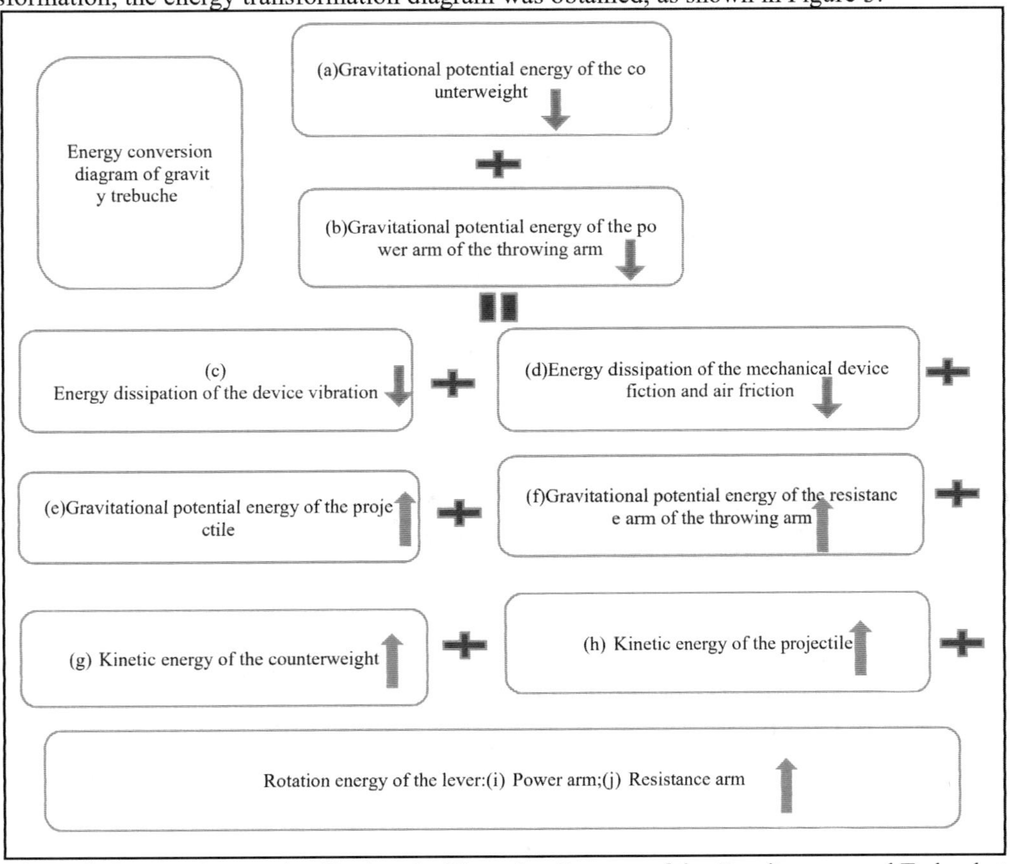

Figure 3. Energy Conversion During the Throwing Process of the Gravity-powered Trebuchet

Graphical analysis was conducted based on the schematic of the throwing process. Let the length of the power arm be L_1, the mass of the power arm be m_1, the length of the resistance arm be L_2 the mass of the resistance arm be m_2, the mass of the counterweight be M, and the mass of the projectile be m, the force diagram of the throwing arm was drawn, and a horizontal line was drawn at the pivot point. When the counterweight is above the horizontal line, the trebuchet is in the energy storage state (also called the pending state). When the counterweight is below the horizontal line, the trebuchet is in a state where the projectile just breaks free from the device (also called the critical state), and its velocity is v. In the two states, the angles between the resistance arm and the horizontal line are θ_1 and θ_2, respectively, and the angle between the velocity of the projectile just breaking free from the device and the horizontal line is θ, as shown in Figure 4.

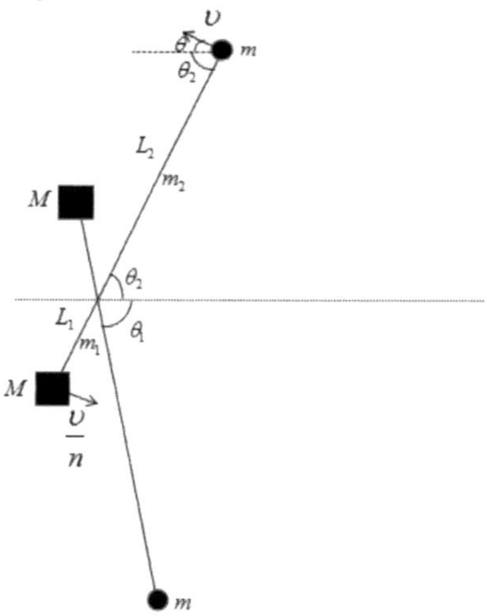

Figure 4. Kinetic Analysis. The dots are projectiles, and the squares are heavy objects.

The energy conversion equation was obtained from the above analysis, as shown in Equation (1) below.

$$MgL_1(\sin\theta_1+\sin\theta_2)(1-15\%)+\frac{m_1gL_1(\sin\theta_1+\sin\theta_2)}{2} = \frac{\mu(M+m+m_1+m_2)(\theta_1+\theta_2)D}{2}+mgL_2(\sin\theta_1+\sin\theta_2) + \frac{m_2gL_2(\sin\theta_1+\sin\theta_2)}{2}+\frac{Mv^2L_1^2}{2L_2^2}+\frac{mv^2}{2}+\frac{m_1L_1^2v^2}{6L_2^2}+\frac{m_2v^2}{6} \quad (1)$$

(Note: In the equation, 15% is the empirical value of vibration energy dissipation.)

In Equation (1), μ represents the friction coefficient, and D represents the nominal inner diameter of the bearing. Different devices have different vibration energy dissipation, and the empirical value was taken here. Generally, the value ranges from 15% to 20% and can be adjusted according to the actual situation, and the deviation is adjustable.

In actual situations, as the power arm of the throwing arm in the gravity-powered trebuchet drops, the resistance arm goes up, so the potential energy change of the arm is very small. Also, it was assumed that the material of the throwing arm is fine and light and has uniform density and the same cross-sectional area. Due to this, the potential energy change of the arm can be ignored. The calculated data showed that the energy dissipated by friction had little effect on the whole device and could be replaced by an empirical coefficient, which is generally 0.97–0.99. The above two are not definite values, and the empirical coefficient depends on the actual situation. Therefore, a simple expression of energy conversion can be obtained, as shown in Equation (2) below.

$$0.98MgL_1(\sin\theta_1+\sin\theta_2)(1-15\%) = mgL_2(\sin\theta_1+\sin\theta_2)+\frac{Mv^2L_1^2}{2L_2^2}+\frac{mv^2}{2}+\frac{m_1L_1^2v^2}{6L_2^2}+\frac{m_2v^2}{6} \quad (2)$$

By simplifying Equation (2), the expression for the velocity can be obtained, as shown in Equation (3) below.

$$v = \sqrt{\frac{6L_2^2(0.833MgL_1 - mgL_2)(\sin\theta_1 + \sin\theta_2)}{3ML_1^2 + 3mL_2^2 + m_1L_1^2 + m_2L_2^2}} \quad (3)$$

3.2 Kinematic analysis of gravity-powered trebuchets

In the first stage of the kinematic analysis of the trebuchet, the velocity of the projectile when it just breaks free from the trebuchet (called off-point velocity) can be calculated. Then, in the second stage, kinematic analysis can be performed for the motion made by the projectile after it just breaks free from the trebuchet from a geometric view. In the throwing process, the motion made by the projectile is oblique projectile motion (In the process, the projectile is only subject to the force of gravity and the influence of air resistance is so small that it can be ignored).

Suppose the height of the pivot point of the trebuchet is H. The projectile breaks free from the device at Point $A(x_0, y_0)$, and the coordinate of the target point is $B(x_p, y_p)$, the projection point of the pivot point on the ground was taken as the origin to establish the coordinate system. The schematic diagram of the motion of the projectile is shown in Figure 5.

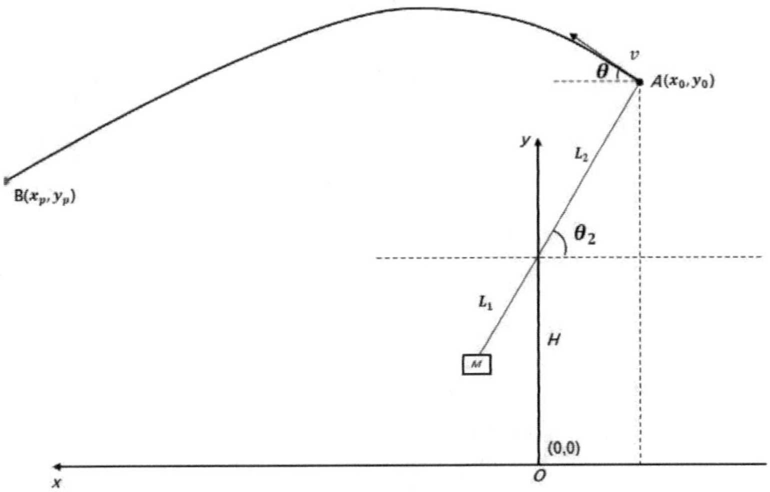

Figure 5. The Projectile Is out of Motion

According to Figure 5 and the geometric relations, we can obtain that

$$\theta = 90^\circ - \theta_2$$
$$x_0 = -L_2\cos\theta_2$$
$$y_0 = H + L_2\sin\theta_2$$

The velocity can be decomposed into a horizontal velocity and a vertical velocity as

$$x = x_0 + v\cos(90^\circ - \theta_2)\,t$$
$$y = y_0 + v\sin(90^\circ - \theta_2)\,t - \frac{gt^2}{2} \quad (4)$$

By combining the two equations in Equation (4) into an equation, we can eliminate t and obtain the trajectory equation of the projectile after it breaks free from the device as follows.

$$y = -\frac{g(x - x_0)^2}{2\sin^2\theta_2\, v^2} + \frac{x - x_0}{\tan\theta_2} + y_0 \quad (5)$$

By substituting the target point $B(x_p, y_p)$ into the trajectory equation and combining Equation (3) for the off-point velocity, the total equation can be obtained as follows.

$$y_p = -\frac{g(x_p + L_2 \cos\theta_2)^2 (3ML_1^2 + 3mL_2^2 + m_1L_1^2 + m_2L_2^2)}{12\,L_2^2 \sin^2\theta_2\,(0.833MgL_1 - mgL_2)(\sin\theta_1 + \sin\theta_2)} + \frac{x_p + L_2\cos\theta_2}{\tan\theta_2} + \sin\theta_2 L_2 + H \quad (6)$$

4. Experimental Verification

4.1. Design of a test gravity-powered trebuchet

At the beginning of the design, the length L_1 and L_2 and the mass m_1 and m_2 of the power arm and resistance arm are measured first, and records are made. The mass M and m of the counterweight and the projectile can be measured first during the assembly process, and records are made. Figure 6 shows the trebuchet and relevant components used in this test plan.

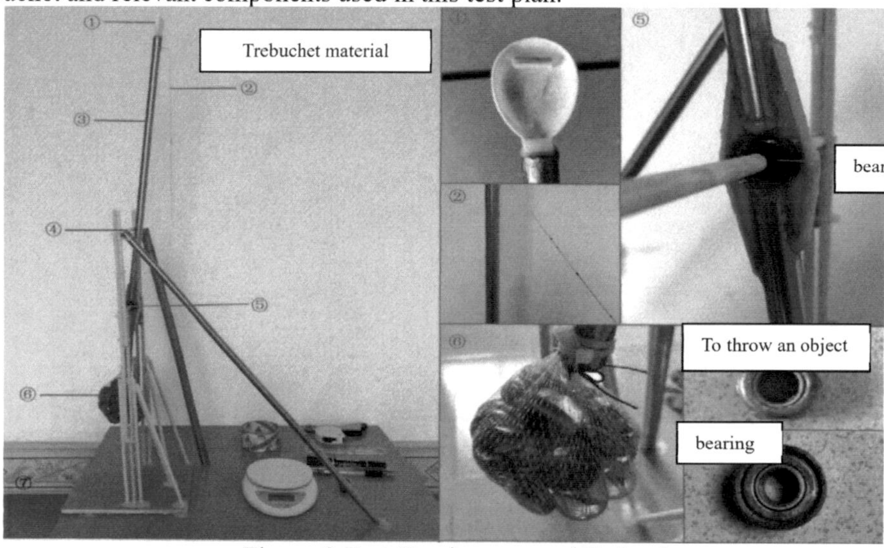

Figure 6. Test Gravity-powered Trebuchet

Materials used in the trebuchet and their roles are as follows.

1) Carrying device of the projectile: a flat plastic round spoon. In order to ensure that the projectile vertically breaks free from the throwing arm in the throwing process, a small wooden bar with a semi-circular cross-section is added at the top of the throwing arm. This ensures the vertical throwing of the projectile.

2) Pulling rope: a light thin rope. The rope is tied to the carrying device of the throwing arm and is mainly used to pull down the throwing arm and mark the throwing corner (see θ_1 marked in Figure 6).

3) Throwing arm: a stainless steel pipe with a diameter of 12 mm, a thickness of 1 mm, and a capacity of 304 L.

4) Crossbeam: disposable chopsticks erected on top of the frame, with a diameter of 5.0 mm and a length of 0.225 m.

5) Bearing: 608 zz deep groove bearings with an inner diameter of 9 mm and an outer diameter of 22 mm.

6) Counterweight: glass bouncing balls. The balls can provide power for the device. You can choose the material of the counterweight at will.

7) Base: a combination of 0.2 m×0.2 m tiles and 0.5 m×0.5 m tiles. The small tiles are used for the base of the frame of the trebuchet, and the large tiles mainly provide space for the frame to be fixed as well as to increase the gravity of the base, so as to improve the stability of the device.

Frame: a combination of disposable chopsticks of 5.0 mm diameter and 0.225 m length and a 304 L stainless steel pipe of 12 mm diameter and 1 mm thickness. Disposable chopsticks are mainly used to form the main frame, while the stainless steel pipe is mainly used to fix the device to reduce the device's vibration energy dissipation.

Projectile: M10 locknut.

Tools used: a utility knife, a needle-nosed plier, a scale, a protractor, a high-precision electronic scale, a hot-melt glue gun and glue stick, thin ropes, etc.

4.2. Design of a test plan

When the device is assembled and placed in the throwing area, the value of its pivot point height H, and the horizontal distance x_p of the selected pivot point from the preset target point are measured. According to the plan, the angle of θ_2 is predetermined first, and generally, an angle from 70° to 85° (depending on the situation) is set to calculate the coordinate point $A(x_0, y_0)$. Then, the angle of θ_1 is set, and the theoretical throwing height y_p is calculated by the total Equation (6). After that, throwing is performed, and each hit point is marked. Finally, the height coordinates of these hit points are measured and recorded as $y_i(i = 1,2, \ldots, i)$, and the arithmetic mean of the hit points is calculated and compared with the theoretical height y_p for analysis, and the sample standard variance of the hit points is calculated to analyze the dispersion of the hit points in each test.

The above test process is repeated by adjusting the angle of θ_1, and the results of each test are analyzed. For the hit rate test, a dartboard is drawn with the target point $B(x_p, y_p)$ as the center of the dartboard, and the radius is as large as possible to ensure that each point can hit the dartboard, and then a target circle (recommended radius: 0.05 m) is drawn on the dartboard with the target point as the center of the circle, as shown in Figure 7.

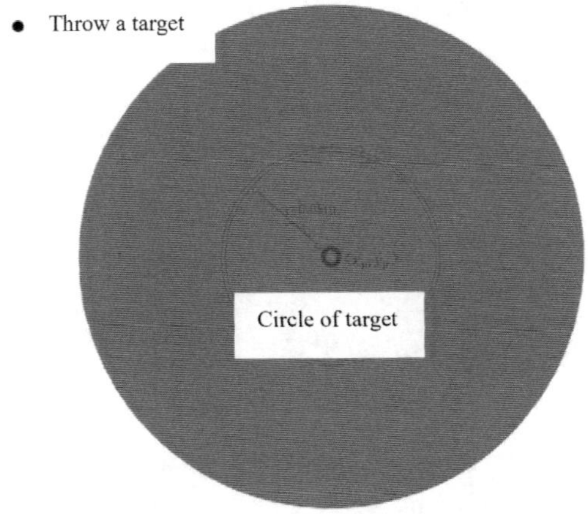

Figure 7. The Design of Target Throwing

The hit rate is equal to the number of times the target circle is hit divided by the total number of casting. Different conditions are formulated, and the hit rate is tested repeatedly. Finally, the results are analyzed.

4.3 Analysis of test results

In this study, several system tests and hit rate tests were conducted, and the on-site throwing mark images of four of these tests are shown in Figure 8.

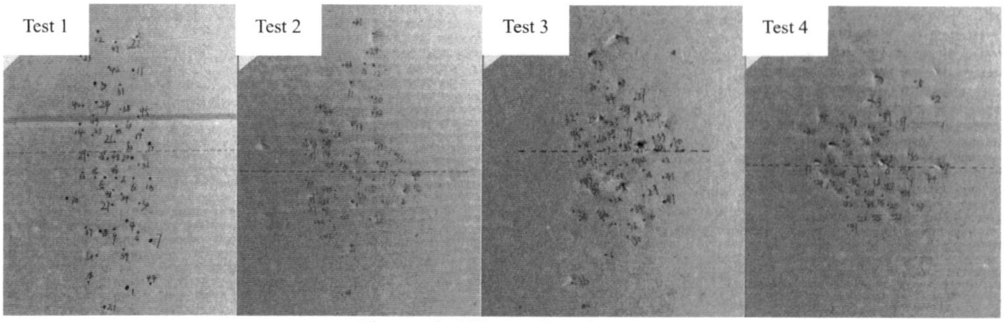

Figure 8. Hit the Dot Mark

Table 1. Experimental Equipment Parameters

Equipment parameters:	
Weight distribution (M):	0.435 kg
Mass of the projectile (m):	0.008 kg
Power arm length (L_1):	0.13 m
Length of resistance arm (L_2):	0.49 m
Power arm mass (m_1):	0.1174 kg
Mass of the resistance arm (m_2):	0.04426 kg
Shaft height (H):	0.36 m
Theta 2:	82°

The data from the four tests were processed as shown in Table 2, and the experimental equipment parameters are shown in Table 1.

Table 2. Test Data Processing

No.	Theta1 (°)	Throwing distance x_p(m)	Theoretical throwing height y_p(m)	The total Number of casting	Best estimated height(m)	Sample standard variance
1	70	2.5	0.147442	50	0.094	0.0325
2	60	2.5	0.105426	47	0.096	0.0301
3	60	2	0.422038	51	0.396	0.0209
4	30	2	0.246707	50	0.202	0.0149

According to the analysis of the results of several tests, theta1 set at 20° to 60° and theta2 set at 75° to 85° are the empirical conditions for the gravity-powered trebuchet to achieve the best throwing performance, which is further verified by several trial throws. This recommendation can be achieved simply by changing other conditions, such as the mass of the counterweight and the ratio of the length of the lever. In the data processing of the hit points, the arithmetic mean of each test was compared with the theoretical value, and the difference between the practical value and the theoretical value was analyzed to be around 0.05 m, which is called the mean error. Since many uncertainties existed in the vibration of the device and the operation of the thrower during the tests, the results fluctuated but not greatly.

In the hit rate tests, the radius of the target circle was set as 0.05 m. The analysis results of the eight tests are shown below. The test conditions are shown in Table 3.

Table 3. Hit Rate Test Condition

No.	Theta1(°)	Theta2(°)	Throwing distance x_p(m)	Theoretical throwing height y_p(m)
1	70	82	2.5	0.147442
2	60	82	2.5	0.105426
3	60	82	2	0.422038
4	40	82	2	0.324454
5	40	82	2	0.324454
6	10	82	2	-0.002613
7	20	82	2	0.141269
8	30	82	2	0.246707

After the data recording and statistical calculation, the average hit rate of these eight tests was obtained as 71%, and the statistics are shown in Table 4.

Table 4. Hit Rate Statistics

No.	Total number of flips	Number of hits in the target circle	Percentage points
1	50	27	54
2	50	38	76
3	50	41	82
4	50	36	72
5	50	42	84
6	50	30	60
7	50	29	58
8	50	40	80
		Average hit rate:0.7075	

5. Conclusion

With a gravity-powered trebuchet as the object of study, this paper established a mechanical model to analyze the trebuchet in stages from the traditional dynamics and kinematics perspectives. With the energy conversion and energy dissipation in practical applications taken into account, the mechanical process of the trebuchet in detail was calculated, the total energy equation was listed, and then the off-point velocity expression was obtained by simplifying the conversion. Then, based on kinematics, the trajectory of the projectile after breaking free from the trebuchet was analyzed, and the equation of motion was listed. Finally, based on the obtained off-point velocity expression, the total equation for the trebuchet was obtained.

Based on the theoretical analysis, a self-designed gravity-powered trebuchet was introduced and used as the main study object of the test plan. Through the practical operation, the feasibility of the theoretical analysis was finally verified, and the performance indicators of the trebuchet were investigated. During the test process, problems were continuously discovered and solved, and then a complete test plan was concluded, which could provide valuable references for future trebuchet tests. Additionally, hit rate tests were conducted for the gravity-powered trebuchet. By combining different conditions of the tests, the mean hit rate was obtained as 71%.

Acknowledgments

This work was supported by the High-level Talents Research Start-up Fund Project of Liupanshui Normal University of China (Contract No. LPSSYKYJJ202203).

References

[1] Chen M. Y. and Jin M. B. Three new perspectives on historical archaeology: Wood and stone

composite weapons—Stone throwing ropes, trebuchets, and catapults [J]. Tribune of Social Sciences, 2014(3): 32–42.
[2] Ji C. Catapults in the Sui and Tang dynasties: form, performance, combat, and dissemination [J]. Tang Shi Lun Cong, 2016(1): 163–178.
[3] Zhang J. B. Research on catapults in ancient China [D]. Henan: Zhengzhou University, 2007. doi:10.7666/d.y1059827.
[4] Chevedden, P. E., Eigenbrod, L., Foley, V. & Soedel, W. 1995. The trebuchet. - Scientific American, July, 58–63.
[5] Hansen, P. V. 1992. Experimental reconstruction of a medieval trebuchet.
[6] Liu C. J. and Zheng J. Q. Dynamics simulation of trebuchet mechanism [J]. Mechanical Engineer, 2007(8):70–72. DOI: 10.3969/j.issn.1002-2333.2007.08.033.
[7] Xiao H. F. Research on a new high-efficiency trebuchet and its innovative design [D]. DOI:10.7666/d.y971180.
[8] Wang X. M. Physics of the operation and projection mechanism of ancient trebuchets[J]. Physics Teacher, 2021,42(12):59–62,66. doi:10.3969/j.issn.1002-042X.2021.12.019.

A General Approach for the Analysis of Transverse Vibration Characteristics of a Rectangular Plate with Arbitrary-shaped Openings

Wenwen Zhang

Institute of Vibration and Sound, Naval University of Engineering, Jiefang Boulevard 717, Wuhan 430033, China

Corresponding author: zwwada@163.com

Abstract. Based on the Rayleigh-Ritz method, the transverse vibration characteristics of a rectangular plate with an arbitrarily shaped inner opening are investigated. A modified two-dimensional Fourier series was chosen as the trial function and springs of different stiffnesses were used to equate the arbitrary boundary conditions of the rectangular plate. The Lagrange generalization of the transverse vibration of the open rectangular plate is obtained by equating the open area to a special film with physical properties of zero according to the energy superposition principle. The variational method is used to derive the characteristic equations of the thin plate and to find the intrinsic frequencies and corresponding vibration modes. The convergence of this theoretical method is analyzed by means of an example, and the accuracy of the method is verified by comparison with the results of finite element software calculations, which provides a reference for practical engineering problems.

1. Introduction

The rectangular plate with an opening is a kind of engineering structure that is widely used in the manufacturing fields of shipbuilding, aerospace, and the automotive industry. In the marine field, in order to reduce the weight of the structure, flooding water, and stealth, openings are often set on panel structures. Typical of these structures are the integrated superstructure of ships, whose plate frame structure has to be opened to accommodate the radar. Since the existence of the opening changes the geometric and physical characteristics of the original plate structure, it will have a significant impact on the transverse vibration characteristics of the plate structure, and the transverse vibration characteristics are one of the important parameters to consider the resonance of power equipment and the structure.

In recent years, scholars have carried out a lot of studies on the vibration problem of rectangular plates. Du et al. [1] employed the two-dimension Fourier series plus the complementary terms to solve the in-plane vibrations of arbitrarily supported rectangular plates. Furthermore, Li et al. [2] extended the equations in Refs. [1] into the transverse vibration case and replaced the supplementary functions in the form of trigonometric functions. Jayasinghe et al. [4] used the Dynamic Coefficient Matrix (DCM) method to obtain solutions for the intrinsic frequencies of rectangular plates and verified the accuracy of this method. Bao et al. [3], on the other hand, combined the spectral geometry expansion and Hamilton's principle to establish a unified two-coordinate model for the analysis of the transverse vibration of rectangular plates.

The analysis of vibration characteristics of rectangular flat plates provides a theoretical basis for the study of transverse vibration of open-ended plates, but the corresponding research results are still relatively few. Currently, the finite element method and the energy method are two common methods used to study the transverse vibration characteristics of plates with openings. Zhang et al. [5] analyzed the vibration characteristics of an open rectangular plate by means of the Hencky Bar-net Expansion Model (HBM). However, the finite element method relies on the user's experience in the meshing process and requires a lot of effort in the accuracy analysis and re-modeling when the model changes. As a common energy method, the Rayleigh-Ritz method is widely used in the study of vibrations of plates with openings because of the simplicity of its solution process. Based on the Rayleigh-Ritz method, Huang et al. [6] used two sets of displacement tolerance functions to investigate the inherent vibration characteristics of a rectangular plate with a V-shaped notch at the edge. Li et al. [7] simplified the vibration solution problem of the plate with an opening by using the energy method and the symmetry relationship of the opening and obtained the intrinsic frequency solution for the free vibration. However, there are still few studies concentrated on the general approach for the transverse vibration of a rectangular plate with an arbitrarily shaped opening.

In this paper, based on the Rayleigh-Ritz method, the two-dimensional Fourier series modified by trigonometric functions is proposed as the trial function of transverse vibration displacement of an open rectangular plate. The opening is regarded as a special film with zero physical properties, and the total energy of the open area is considered to be zero. The Lagrange equation for the open plate is obtained by subtracting the energy of the original plate domain at the opening from the total energy of the unopened plate and solving the vibration energy equation based on the variational method. This leads to a general method of solving bending vibrations for rectangular plates with arbitrary openings.

2. Theoretical Solution Process

The rectangular plate with an opening in this paper is considered to satisfy Kirchhoff's thin plate theory, and its transverse displacement functions are expressed by the modal summation approach. The boundary conditions are set by the combination of elastic deflectional restraint and elastic rotation restraint, as shown in Figure 1. Accordingly, any boundary condition can be equivalent by adjusting the deflection restraint stiffness K_d and the rotation restraint stiffness K_r. For example, K_d =0 and K_r =0 represent the free edge, K_d =∞ and K_r =0 represent the simply supported edge, and K_d =∞ and K_r =∞ represent the clamped edge.

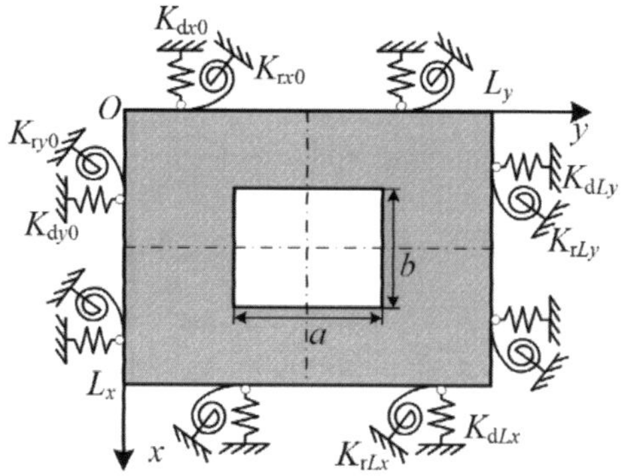

Figure 1. The Rectangular Plate with An Opening under Arbitrary Boundary Conditions Equivalented by Elastic Restraints

For the plate satisfies Kirchhoff's thin plate theory, the bending vibration of the plate is governed by,

$$D\nabla^4 w(x,y) - \rho_s h\omega^2 w(x,y) = f(x,y) \tag{1}$$

where $\nabla^4 = \partial^4/\partial x^4 + 2\partial^4/\partial x^2\partial y^2 + \partial^4/\partial y^4$; $w(x,y)$ is the flexural displacement of the plate; ω is the angular frequency; $D = \dfrac{Eh^2}{12(1-\sigma^2)}$ is the bending stiffness, E and σ are Young's modulus and Poisson's ratio, respectively; ρ_s and h are the density and thickness of the plate, respectively; $f(x,y)$ is the external excitation acting on the plate surface. The time-dependent term $e^{j\omega t}$ is omitted for simplicity.

The boundary conditions of a rectangular plate with elastically restrained edges have been given in Refs. [2]. The governing Equation (1) for rectangular plates with a set of elastic boundary conditions cannot be exactly solved. Consequently, the Rayleigh-Ritz method is employed in this study to find an approximate solution. The trial function of flexural displacement for arbitrarily supported plates is chosen in the form of supplemented two-dimensional Fourier cosine series, as follows:

$$\begin{aligned} w(x,y) = &\sum_{m=0}^{\infty}\sum_{n=0}^{\infty} A_{mn}\cos(\alpha_m x)\cos(\beta_n y) + \sum_{m=0}^{\infty}\sum_{n=1}^{4} B_{mn}\cos(\alpha_m x)\sin(\beta_n y) + \\ &\sum_{m=1}^{4}\sum_{n=0}^{\infty} C_{mn}\sin(\alpha_m x)\cos(\beta_n y) + \sum_{m=1}^{4}\sum_{n=1}^{4} D_{mn}\sin(\alpha_m x)\sin(\beta_n y) \end{aligned} \tag{2}$$

where $\alpha_m = m\pi/L_x$; $\beta_n = n\pi/L_y$; L_x and L_y are the dimensions of the rectangular plate in the x and y directions, respectively; A_{mn}, B_{mn}, C_{mn}, and D_{mn} are the corresponding expansion coefficients. It can be easily verified that the displacement function shown in Equation (2) naturally satisfies arbitrary boundary conditions by setting appropriate expansion coefficients. The supplementary terms, which contain the sine function, guarantee the displacement function to be sufficiently smooth; that is, the third-order derivatives are continuous, and the fourth-order derivatives exist over the entire plate domain.

According to the energy method, the Lagrange function $L(w)$ of the transverse vibration of a rectangular plate with an opening can be expressed as a function of the total potential energy $V(w)$, the total kinetic energy $T(w)$, and the work done by the external excitation $Q(w)$, as

$$L(w) = V(w) - T(w) - Q(w) \tag{3}$$

The opening of the plate can be considered a special film with zero physical properties, zero material density, Young's modulus, and Poisson's ratio. Therefore, the energy equation of the rectangular plate with an opening can be obtained by subtracting the energy of the opening domain from the energy of a rectangular plate with no opening. For the rectangular plate with no opening, the potential energy, kinetic energy, and external work done can be, respectively, expressed as

$$\begin{aligned} V_{p0}(w) = &\frac{D}{2}\int_0^{L_x}\int_0^{L_y}\left[\left(\frac{\partial^2 w}{\partial x^2}\right)^2 + \left(\frac{\partial^2 w}{\partial y^2}\right)^2 + 2\sigma\frac{\partial^2 w}{\partial x^2}\frac{\partial^2 w}{\partial y^2} + 2(1-\sigma)\left(\frac{\partial^2 w}{\partial x\partial y}\right)^2\right]\mathrm{d}x\mathrm{d}y \\ &+\frac{1}{2}\int_0^{L_x}\left[K_{dy0}w^2 + K_{ry0}\left(\partial w/\partial y\right)^2\right]_{y=0}\mathrm{d}x + \frac{1}{2}\int_0^{L_x}\left[K_{dyb}w^2 + K_{ryb}\left(\partial w/\partial y\right)^2\right]_{y=L_y}\mathrm{d}x \\ &+\frac{1}{2}\int_0^{L_y}\left[K_{dx0}w^2 + K_{rx0}\left(\partial w/\partial y\right)^2\right]_{x=0}\mathrm{d}y + \frac{1}{2}\int_0^{L_y}\left[K_{dxa}w^2 + K_{rxa}\left(\partial w/\partial x\right)^2\right]_{x=L_x}\mathrm{d}y \end{aligned} \tag{4}$$

$$T_{p0}(w) = \frac{1}{2}\rho_s h\omega^2\int_0^{L_x}\int_0^{L_y} w^2\mathrm{d}x\mathrm{d}y \tag{5}$$

and

$$Q_{p0}(w) = \int_0^{L_x}\int_0^{L_y} f(x,y)w(x,y)\mathrm{d}x\mathrm{d}y \tag{6}$$

The energy of the plate at the opening can be expressed as:

$$V_{p1}(w) = \frac{D}{2}\iint_S\left[\left(\frac{\partial^2 w}{\partial x^2}\right)^2 + \left(\frac{\partial^2 w}{\partial y^2}\right)^2 + 2\sigma\frac{\partial^2 w}{\partial x^2}\frac{\partial^2 w}{\partial y^2} + 2(1-\sigma)\left(\frac{\partial^2 w}{\partial x\partial y}\right)^2\right]\mathrm{d}x\mathrm{d}y \tag{7}$$

$$T_{p1}(w) = \frac{1}{2}\rho_s h\omega^2\iint_S w^2\mathrm{d}x\mathrm{d}y \tag{8}$$

and

$$Q_{p1}(w) = \iint_S f(x,y)w(x,y)\mathrm{d}x\mathrm{d}y \tag{9}$$

where the integral subscript S indicates the open domain. For the central rectangular opening, as shown in Figure 1, the surface integral $\iint_S$ can be written as $\int_{(L_y-b)/2}^{(L_y+b)/2}\int_{(L_x-a)/2}^{(L_x+a)/2}$, and a and b are the length and width of the rectangular opening, respectively.

According to Hamilton's theorem, by substituting Equations (4)– (9) into the Lagrange function Equation (3) and taking extreme values for each of the expansion coefficients, we can obtain the system of equations to solve the expansion coefficient, as

$$\frac{\partial L(w)}{\partial A_{mn}} = \frac{\partial\left\{V_{\mathrm{p0}}(w)-V_{\mathrm{p1}}(w)-\left[T_{\mathrm{p0}}(w)-T_{\mathrm{p1}}(w)\right]-\left[Q_{\mathrm{p0}}(w)-Q_{\mathrm{p1}}(w)\right]\right\}}{\partial A_{mn}} = 0 \tag{10}$$

The other coefficients, B_{mn}, C_{mn}, and D_{mn}, have the same partial derivative relationship as the equation above.

Substituting Equations (4)– (9) into Equation (10), the expansion coefficients solution function can be finally rewritten in a matrix form as

$$(\boldsymbol{K}-\rho_s h\omega^2\boldsymbol{M})\boldsymbol{P}=\boldsymbol{F} \tag{11}$$

where $\boldsymbol{P}=[A_{00},\cdots,A_{MN},B_{01},\cdots,B_{M4},C_{10},\cdots,C_{4N},D_{11},\cdots,D_{44}]^{\mathrm{T}}$ is the vector of expansion coefficients; M and N are the modal truncation numbers of m and n, respectively; $\boldsymbol{F}$ is the vector of excitation force; $\boldsymbol{K}$ and $\boldsymbol{M}$ are the stiffness matrix and mass matrix, respectively. It should be noted that $\boldsymbol{K}$ and $\boldsymbol{M}$ are symmetric matrices.

For a given force, Equation (11) can be directly solved for the expansion coefficients and hence, the plate displacement. By setting the force vector to zero, Equation (11) will reduce to a standard characteristic equation from which the free vibration properties of the elastically restrained plate can be readily calculated.

3. Numerical Calculation and Analysis

The geometry of the rectangular plate with an opening in this analysis example is chosen to have a central rectangular opening, as shown in Figure.1. The geometry parameters of the analyzed rectangular plate with an opening are L_x=0.32 m, L_y=0.2 m, a=0.09 m, b=0.08 m, and h=0.0025 m. The mechanical parameters of the open rectangular plate refer to stainless steel, with ρ_s=7580 kg/m^3, Poisson's ratio σ=0.3, and Young's modulus E=21.6 GPa.

3.1. Modal truncation number convergence analysis

When the modal superposition method is used to approximate the vibration displacement of the elastic plate by the Fourier series, the modal truncation numbers M and N greatly impact the accuracy of the calculation results. Since the increase of M and N will also enhance the calculation burden, it is necessary to analyze the convergence of the modal truncation numbers M and N. The boundary conditions of the rectangular plate with an opening are set to be clamped on four sides (C-C-C-C) for analysis. The clamped boundary condition is achieved by setting the deflectional stiffness K_{d} and rotational stiffness K_{r} to large enough values. Here, we set K_{d} =10^{12} N/m^2 and K_{r} =10^{12} N/rad, respectively. In order to simplify the analysis, the x-directional modal truncation number M and y-directional modal truncation number N were taken to be the same value. The first 6 orders of the inherent frequency of the rectangular plate with an opening are calculated by the approach in this paper, as shown in Table 1.

Table 1. Modal Frequencies of the C-C-C-C Type Boundary Plate with Central Rectangular Aperture

M=N	Modal order / Hz					
	1	2	3	4	5	6
14	468	549	870	1 035	1 079	1 327
16	467	545	869	1 030	1 079	1 321
18	467	543	869	1 028	1 079	1 317
20	466	542	868	1 027	1 078	1 315
22	465	541	868	1 027	1 078	1 314
24	465	540	868	1 026	1 078	1 313
26	465	540	868	1 026	1 078	1 312
28	465	540	868	1026	1 078	1 312

As can be seen from Table 1, with the increase of the truncation numbers M and N, the first 6 orders of modal frequencies of the open rectangular plate converge rapidly. At a modal truncation number of M=N=26, the first 6 orders of inherent frequencies of the rectangular cavity with an opening remain basically unchanged. Compared to the transverse displacement trail functions of the rectangular plate with an opening proposed in Ref. [7] and Ref. [8], the improved dual Fourier series test function proposed in this paper has better convergence. In the later examples, the modal truncation number for the transverse vibration calculations of the rectangular plate with an opening is taken as M=N=26.

3.2. Boundary stiffness convergence analysis

As mentioned above, in the theoretical approach proposed in this paper, the boundary stiffness is set as 0 or ∞, which can be equivalent to classical boundary conditions such as clamped support and simple and free support. However, in actual calculations, the boundary stiffness of elastic constraint must be replaced by a specific value, so it is necessary to conduct a convergence analysis of the influence of boundary stiffness on the transverse vibration of the rectangular plate with an opening. The free vibration of the rectangular plate with an opening was also taken as the analysis object, and the parameters were set in the same section as above. The deflectional stiffness K_d and rotational stiffness K_r of the four edges were kept unchanged respectively, and their values were fixed as $K_d = 10^{12}$N/m^2 and $K_r = 10^{12}$N/rad respectively, and the K_r and K_d of the four edges were adjusted accordingly. The variation curves of the first three intrinsic frequencies of the opened rectangular plate versus the boundary deflectional stiffness K_d and the rotational stiffness K_r were calculated by the approach proposed in this paper, as shown in Figure 2.

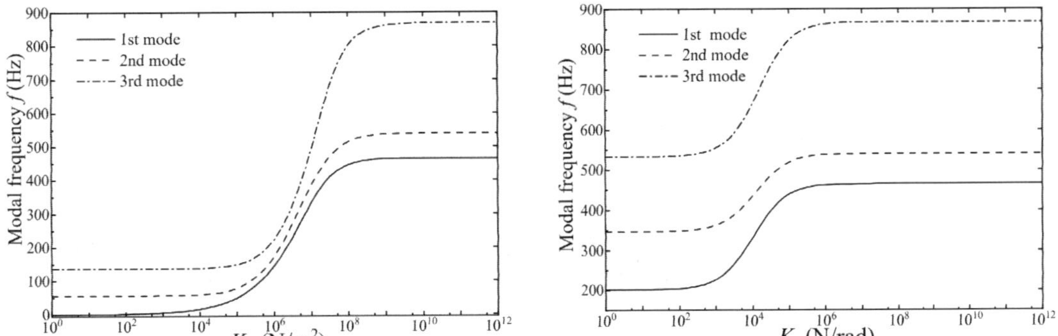

(a) Variation of deflectional stiffness (b) Variation of rotational stiffness

Figure 2. The First 3 Modal Frequencies Varied with the Boundary Stiffness

From Figure 2, it can be seen that when the deflectional stiffness K_d is greater than 10^{10}N/m^2, the effect of the deflectional stiffness on the vibration of the rectangular plate with an opening has basically converged and can be regarded as a deflectional rigid constraint, and when K_r is greater than 10^7 N/rad, the effect of the boundary corner stiffness on the rectangular plate with an opening has basically converged and can be regarded as a rotational rigid constraint. Therefore, it is reasonable to use $K_d=10^{12}$ N/m^2 and $K_r=10^{12}$ N/rad as the deflectional and rotational rigid boundary conditions respectively in the previous paper.

3.3. Accuracy analysis

The commercial finite element software COMSOL is employed in this section to verify the accuracy of the proposed theoretical approach on calculating the transverse vibration of the rectangular plate with an opening. The finite element number of the open rectangular plate for calculating the transverse vibration is taken as 160×100 with the C-C-C-C type boundary condition. The first 10-order modal frequencies obtained from the proposed theoretical approach and the FEM method are shown in Table 2.

Table 2. Comparisons of Theoretical Results and FEM Results of the Modal Frequencies for the Opened Rectangular Plate with the C-C-C-C Type Boundary Condition

Modal order	Theoretical results/Hz	FEM result/Hz	Error/%
1	464.7	468.3	0.77
2	540.0	548.6	1.59
3	868.3	875.4	0.82
4	1025.7	1030.7	0.49
5	1077.9	1085.6	0.71
6	1311.6	1321.8	0.78
7	1372.1	1390.5	1.34
8	1618.1	1632.2	0.87
9	1808.7	1822.5	0.76
10	1906.4	1924.8	0.97

As can be seen from Table 2, the theoretical modal frequency results and the FEM results are in good agreement. The maximum deviation of the first 10 orders of modal frequencies for the opened rectangular plate is only 1.59%, indicating that the theoretical calculation of the proposed approach is highly accurate. Further, the theoretical and FEM results of the 1st, 3rd, and 6th-order mode shapes of the rectangular plate with an opening are compared, as shown in Figure 3. It can be seen that the theoretical and FEM results of the above orders of mode shapes are in high agreement, which verifies the high accuracy of the proposed approach on calculating the transverse vibration of the rectangular plate with an opening.

Theoretical results FEM results

(a) 1st mode shape

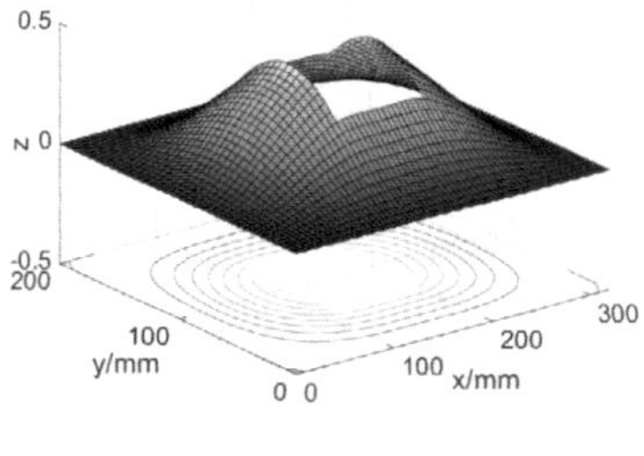

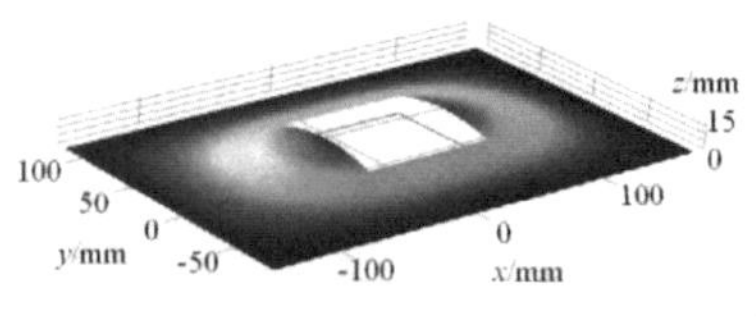

(b) 3rd modal shape

(c) 6th mode shape

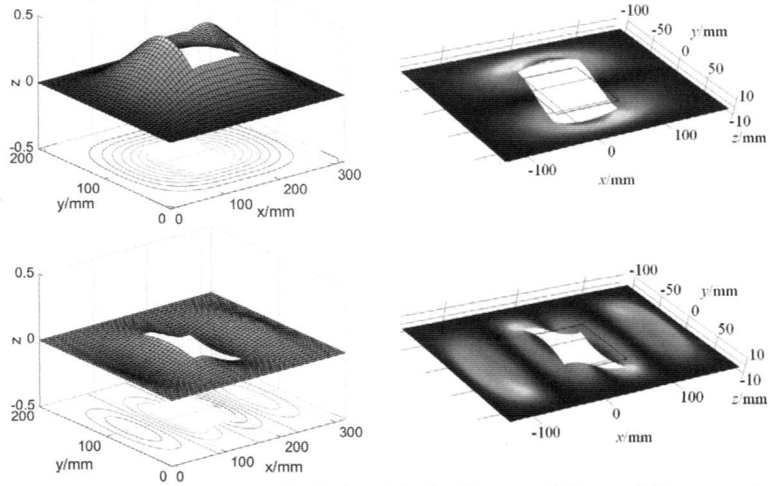

Figure 3. Comparison of Theoretical and FEM Calculation Mode Shapes of Opened Rectangular Plates under the C-C-C-C Type Boundary Condition

4. Conclusion

This study analyzes the free transverse vibration characteristics of a rectangular plate with arbitrarily shaped openings based on the Rayleigh-Ritz method. The approach proposed in this paper uses a modified two-dimensional Fourier series as a trial function for the transverse vibration displacement to calculate the strain energy and kinetic energy of the structure. A spring boundary model is used to simulate the arbitrary boundary conditions of the rectangular plate to calculate its spring elastic potential energy. Finally, the controlling equations of the free transverse vibration of the open rectangular plate are solved using the energy-generalized variational principle to obtain its natural frequency and vibration characteristics. The algorithm shows that this proposed approach has good convergence and accuracy, and it is applicable to rectangular plates with arbitrary openings.

References

[1] Du J, Li W L, Jin G, Yang T and Liu Z 2007 *J. Sound Vib.* **306** 908–27
[2] Li W L, Zhang X, Du J and Liu Z 2009 *J. Sound Vib.* **321**, 254–69
[3] Bao S and Wang S 2019 *Arch. Appl. Mech* **89**(8) 1485–99
[4] Jayasinghe S and Hashemi S M 2018 *Shock Vib.* **2018** 1–8
[5] Zhang Y P, Wang C M and Pedroso D M 2018 *J. Sound Vib.* **432** 65–87
[6] Huang C S, Leissa A W and Liao S C 2008 *J. Mech. Sci.* **50**(8) 1255–62
[7] Li K, He S, Wu G, Mao Y and Li T 2017 *J. Vib. Shock* **36**(11) 161–5 (in Chinese)
[8] Lu B, Chen Y, Feng Z, Zhang G, Yan W and Xu Q 2020 *J. Vib. Shock* **39**(2) 178–87 (in Chinese)

Research on 3D Modeling and Dynamic characteristics of wind turbine blade

Pengpeng Li[1,2], Gulbahar • Tohti[1,2*], Mamtimin • Geni[1,2*], Wensong Ma[1,2]

1.School of Mechanical Engineering, Xinjiang University, Urumqi, Xinjiang 830047,China; 2.Company of Baibo Electromechanical Technology, Urumqi, Xinjiang 830011,China

Email:1575735082@qq.com;gulbahart@163.com.

Abstract: In order to prevent blade resonance, reduce blade deflection, improve wind turbine power generation efficiency and wind energy utilization rate, the blade model of 3 kW wind turbine is established, and the natural frequency of wind turbine blade is analyzed. Based on the vibration theory, the effects of dynamic stiffening and rotational softening on the natural frequencies of blades are analyzed. Study the change rule of blade vibration mode under rotating speed of 0~300rpm, and analyze the influence of two effects on blade frequency. The results show that at low speed, the rotating softening effect of the blade is obvious, and the vibration frequency decreases; At high speed, the stress stiffening effect is more obvious, the vibration frequency increases, and the coupling vibration has greater destructiveness.

1.Introduction

As a kind of green energy, wind energy has been recognized by countries all over the world. In recent years, wind power generation has become an important direction of world energy development[1,2].With the rapid development of wind turbines, damage accidents of wind turbines gradually increase, among which blade damage leads to wind turbine damage is relatively common[3].Most wind farms are located in remote and harsh areas, and damage to wind turbines will increase operation and maintenance costs. Therefore, it is very important to study the vibration characteristics of fan blades for the development of wind power industry.

In recent years, scholars at home and abroad have done a lot of research work on vibration characteristics and structural characteristics of wind turbine blades. Murtagh et al. [4] used Ansys software to analyze wind turbine blades for static analysis, and put the centrifugal force obtained by it into the stiffness matrix in the form of prestress to obtain the result that the vibration mode frequency of the blades changes with the speed. Junkai Guo et al.[5] carried out finite element analysis on the dynamic characteristics of 300W wind turbine blades. The research shows that the centrifugal force will cause the blade to be subjected to the stress hardening effect, which will increase the natural frequency of the blade. Lijun Shan et al.[6] conducted finite element analysis on 3MW wind turbine blades to obtain blade frequency, and used multi-objective genetic algorithm to optimize blade design, providing solutions to the fan blade resonance problem. Jundong Huang et al.[7] adopted eigenvalue method to analyze the structural characteristics and aerodynamic characteristics of large swept back blades, the structural characteristics and flutter characteristics of the blade are studied, and the influence of blade sweep back on its bending torsional coupling deformation is revealed. Guoyu Hu et al.[8] carried out finite element calculation for 5MW fan blades, and the research results show that stress stiffening has an impact on the natural frequency of blades. Xu et al.[9] analyzed the vibration characteristics of aeroengine blades by using the Ansys finite element analysis method, and proved that the vibration mode and frequency characteristics of blades are quite different at different speeds. Rao et al.[10-11] studied the vibration characteristics of turbine blades, calculated the centrifugal force of the blades using finite element Ansys, and proved that the blades are prone to resonance at low order. Chen et al.[12] studied the complex curvature mode shapes and modal coupling dynamics of wind turbines based on numerical simulation. Xiangming Jin et al.[13] proposed a vibration analysis method based on a centrifugal impeller blade. By measuring the variance of blade thickness, the thickness is adjusted to adjust the natural frequency of the blade, so as to avoid resonance. Xueshi Yao et al. [14] studied the vibration characteristics of blades, and the results showed that the coupling vibration generated by different rotation speeds and vibration frequencies was more destructive

Therefore, on the basis of the above studies, this paper proposes a wind turbine blade dynamics research method based on perturbation modal analysis for small wind turbine blades, and uses Ansys Workbench platform to analyze blades. Firstly, the mechanism of rotational softening effect and stress hardening effect of blades were studied. A 3D finite element model of a 3kw blade was established, and the effects of rotational softening and stress rigidness at 0-300r/min were studied by numerical simulation, and the variation rule of vibration characteristics of the blade was obtained.

2.Methods and Materials

Modal analysis is mainly used to determine the design of blade structure or fan blade vibration characteristics, such as wind turbine blade frequency and mode. In structural design, modal analysis is one of the very important links, and also the early foundation of dynamic analysis. In classical mechanics, the vibration mechanical equation of an object is:

$$[M]\{\ddot{u}\}+[C]\{\dot{u}\}+[K]\{u\}=\{F(t)\} \quad (1)$$

Type: $[M]$ is the quality matrix; $[C]$ is the damping matrix; $[K]$ is the stiffness matrix; $\{\ddot{u}\}$, $\{\dot{u}\}$, $\{u\}$ is the acceleration vector, velocity vector and displacement vector of the node, $\{F(t)\}$ is the external load.

The blade root position of the fan blade is relatively thick, the edge is relatively thin, long and thin, and these characteristics of the fan blade are easy to have bending vibration, torsional vibration or the combination of bending vibration and torsional vibration. The main failure form of the fan blade is the fracture caused by the resonance between the inherent frequency of the fan blade structure and the external excitation. Therefore, in order to avoid the coincidence of the inherent frequency of the blade and the external excitation, it is necessary to carry out the modal analysis of the fan blade.

3.Finite element analysis

The mechanical properties of nylon fiber reinforced composites are orthotropic. In this paper, the research object is a small fan blade, and the blade airfoil is NACA4412.The blade length is 1.6m, the diameter of the wind wheel is 3.3m, the rated power is 3kw, and the speed is 300r/min. Material properties The wind turbine blade is made of nylon fiber reinforced composite material. Show table 1 for detailed material parameters of fan blades.

Table 1 Performance Parameters of Nylon Fiber Reinforced Composite

Density（kg/m^3）	Young's modulus(MPa)	Poisson's ratio
1250	8300	0.28

3.1Establishment of 3D model of blade

The surface of the fan blade is an irregular surface, and part of the detailed surface information will be lost if the three-dimensional software is directly used for modeling. Therefore, the German ATOS optical scanner is used to scan the blade and obtain the point cloud data of the blade. The obtained point cloud data of the blade is imported into the Imageware software for corresponding processing to generate the surface. The obtained point cloud data was imported into GeomagicdesignX to optimize the point cloud data by deleting isolated points, unified sampling, noise reduction, polygon encapsulation, feature removal, hole filling, etc., to obtain accurate point cloud data, and repair the leaf model. The model is repaired and imported into Solidworks to view the model and guide the model into x_t file format, which is convenient for the finite element analysis of the model.

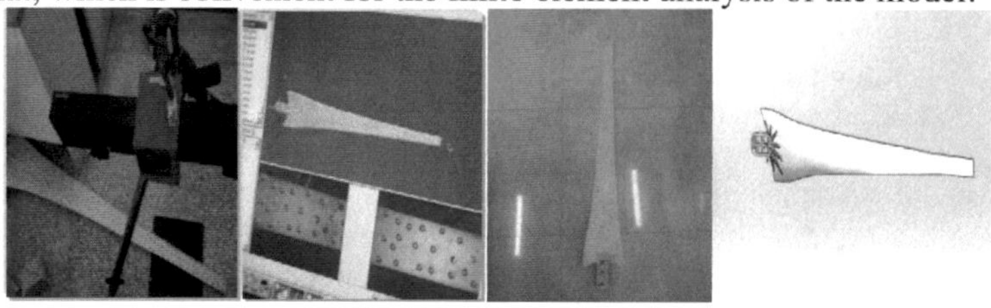

Fig.1 3D blade model

3.2Finite element analysis of fan blades

The static analysis of the model in the finite element analysis can be divided into four steps, that is, the introduction of the blade solid model, the setting of material parameters, the meshing and the setting of boundary conditions.

3.2.1 Blade meshing and adding constraints

The Workbench in ANSYS software is used to grid the fan blade. The small fan blade studied in this paper is a solid structure, which needs to consider the characteristics of nylon fiber composite materials. The model uses the entity solid186 unit, where it is not necessary to define the unit real constant. Use the meshing tool in the Workbench to select the solid cells and Sizing in the MeshControl module to define the grid size as 8mm, the number of cells as 35649, and the number of nodes as 65261.See Figure for finite element blade model.

3.2.2Boundary condition loading

In this study, blade boundary constraints were simplified as follows: the root of the blade and the fixed end of the hub could be regarded as fixed support connection, and the tip and middle of the blade were free ends. In ANSYS finite element model, the three pores at the root position of the fan blade can be fixed constraints.

3.2.3The solution result

According to the finite element model created above, the 10th-order natural frequency of a single blade of a wind turbine is calculated under the static state. The BlockLanczos solution is used to obtain the inherent characteristics of a single blade.

Table 2 Vibration mode of blade

Order	1st	2nd	3rd	4th	5th	6th
Frequency（Hz）	3.77	12.55	30.17	35.56	53.48	81.96

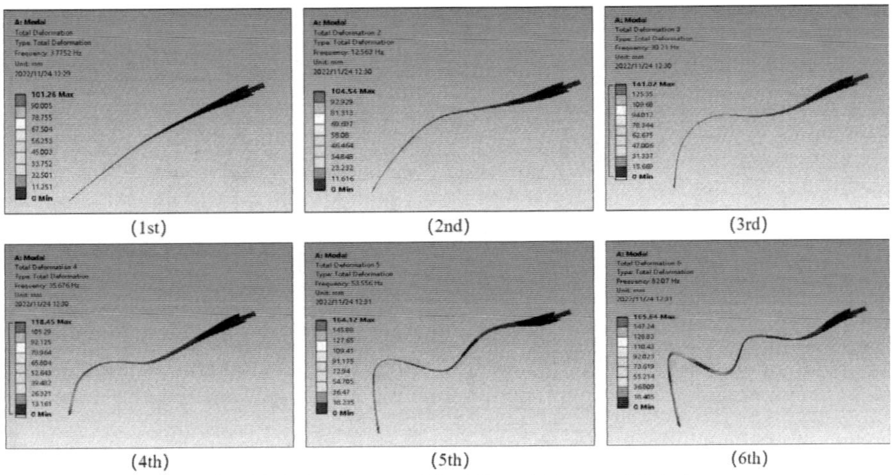

Fig.2 Mode Array of the First 6 Steps of Blade

4.softening and stress hardening effect

4.1Rotary softening

During the operation of wind turbine blade, the fan blade rotates in a wide range in space at a certain angular speed. Under the action of centrifugal force, the natural frequency of the fan blade will change, resulting in the deformation of the blade, and finally the stiffness of the blade will be reduced. This phenomenon is called rotational softening.

When the wind turbine blade is rotating, it will be affected by centrifugal force. When the centrifugal force is too large, the longitudinal deformation will be impossible to ignore. If the longitudinal deformation is large, it will affect the stiffness of the blade. The following simple model is used to illustrate the rotational softening.

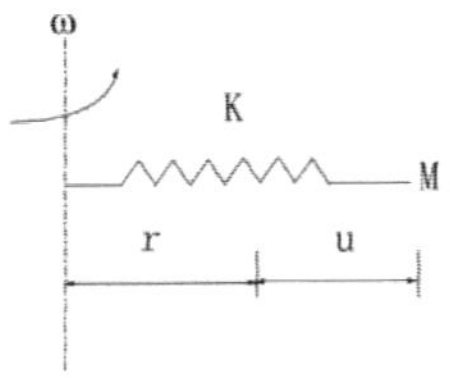

Fig.3 Spring mass rotation system

When the deformation is small and u is small enough, its deformation can be ignored and its vibration equation can be written as

$$M\ddot{u} + Ku = 0 \tag{2}$$

Its equilibrium equation can be expressed as

$$\mathrm{K}u = M\omega^2 \mathrm{r} \tag{3}$$

Where, K is the spring stiffness; U for the displacement of a particle's position relatively free, M is the quality of the particle, ω for the structure of the rotating angular velocity, r for particle free the position of axis relative to the distance.

When the deformation is relatively large, u cannot be ignored, and the corresponding vibration equation of the system can be derived as

$$M\ddot{u} + \mathrm{K}u = \mathrm{f(t)} \tag{4}$$

The equilibrium equation of the system becomes

$$\left(K - M\omega^2\right)u = M\omega^2 \mathrm{r} \tag{5}$$

Take $\overline{K} = K - M\omega^2$

And we can say, $\overline{K} = K - M\omega^2 \leq K$, Thus, the stiffness of the blade is reduced in the process of rotation, and the rotational softening effect is generated.

4.2Hardening of stress

The mechanical mechanism of stress hardening effect can be explained by taking beam element as an example [15] :

Due to the coupling of motion and deformation, F generates additional torque M:

$$M = F\frac{\partial y}{\partial x}dx \tag{6}$$

Where F is the axial force and M is the bending moment。

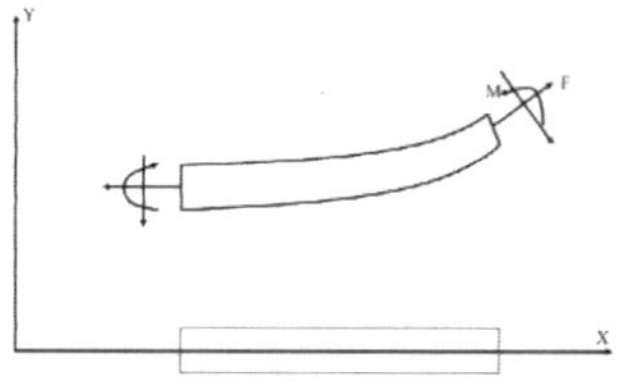

Fig.4 Beam Element Model

In the real process of motion, the coupling of motion and deformation will lead to the additional bending moment M generated by the axial force. When the axial force becomes large, the additional bending moment cannot be ignored. In the rotating body (fan blade), the axial force provides the rotational centrifugal force for its rotation. Additional stiffness occurs, resulting in a dynamic hardening effect.

4.3 Perturbation mode analysis

Before the modal analysis, the static analysis is used to calculate the stress generated by the rotation of the fan blade, and then the stress is applied as the prestress in the modal analysis.

The wind turbine blades were fixed to move and rotate in other directions, and a rotation speed was applied to rotate around the Z-axis, with the rotation speed ranging from 0 to 300r/min. When the blade rotation speed was selected as 10,50,130,220, and 300r/min,the influences of stress rigid effect on the inherent characteristics of the blades were respectively considered. The following 6 groups of data can be obtained.The calculation results of each order of natural frequency at different rotational speeds are shown in Table 3 and 4.

Table 3 Vibration mode of blade based on prestress

Speed /r·min	1st	2nd	3rd	4th	5th	6th
10	3.78	12.56	30.17	35.57	53.49	81.97
50	3.94	12.71	30.27	36.02	53.63	82.12
130	4.77	13.56	30.81	36.91	54.45	83.00
220	6.12	15.26	31.78	36.87	56.23	84.89
300	7.46	17.22	32.72	38.15	58.49	87.28

Table 4 Vibration mode of blade based on spin softening

Speed r/min	1st	2nd	3rd	4th	5th	6th
10	3.76	12.56	30.17	35.56	53.49	81.97
50	3.45	12.63	30.12	35.53	53.62	82.10
130	2.75	12.99	30.37	35.52	54.41	82.96
220	2.38	13.55	31.19	35.87	56.15	84.83
300	2.22	13.98	31.94	36.85	58.42	87.24

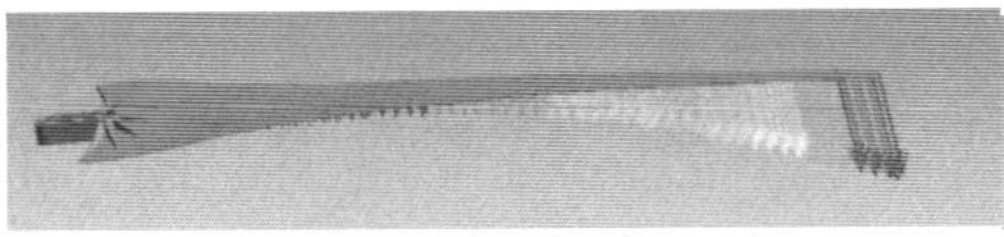

Fig.5 The first flapping vibration

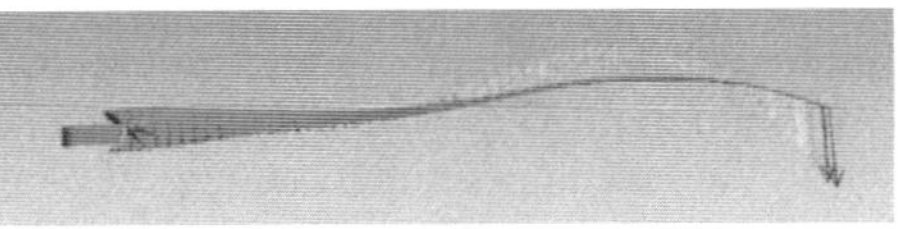

Fig.6 The first fluttering vibration

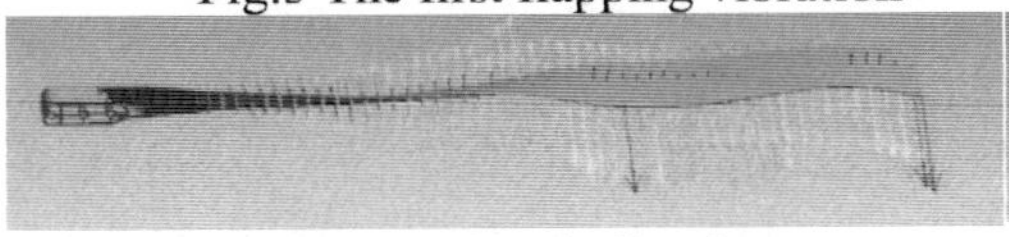

Fig.7 The first torsional vibration

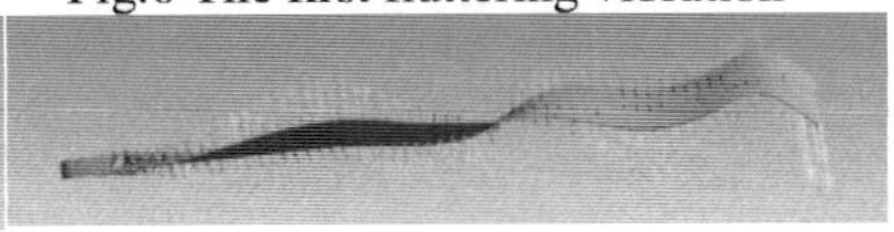

Fig.8 Coupled vibration

The different vibration modes of the blade are shown in FIG.4 to FIG.7, where x-y plane is the blade plane and z is the axis of rotation. With the increase of blade rotational speed, when the blade rotational speed is close to 300r/min, coupling vibration of torsional and flapping occurs. The vibration modes are shown in Fig7 Other forms of coupled vibrations between flapping, pendulums and torsional vibrations occur as the velocity changes.

5.Conclusion

Based on the three-dimensional model of wind turbine blades established by our research group and the basic theory of vibration analysis, a vibration analysis method of rotating blades based on perturbation mode analysis is proposed in this paper. In this paper, the rotary softening effect and stress hardening effect of 1.6 meter wind turbine blades are studied, and a conclusion can be drawn:

(1)With the increase of the modal order of the blade, the blade mode shapes are mainly brandishing and swinging in the low order, and the coupling of brandishing, swinging and torsion appears in the mode shapes above the 7th class. When the critical speed of the blade speed is greater than the normal speed of the blade, the resonance failure will not occur, but with the increase of the order, the vibration displacement will be intensified after the fifth order, and the fatigue fracture is easy to occur.

(2)Both the stress hardening effect and the rotary softening effect have certain influence on the natural frequency of the blade. According to the analysis at 0~300r/min, the rotational softening effect is more obvious when the blade is rotating at a low speed, and the stress hardening effect is more influential when the blade is rotating at a high speed, and the pendulum vibration mode is more susceptible to the rotational softening effect. The effects of stress hardening and rotation softening on large blades are more obvious, so it becomes more important to analyze the effects of both on blades in service.

References:

[1]Xiaorong Wang, Weisheng Wang, Huizhu Dai. Current situation and Prospect of Wind power generation. China Electric Power, 2004,37 (1):81-84
[2]Hongli Zhou, Hong Wang, Luo Zhen, et al. Research Progress of Composite Blades for Wind Power Generation. Materials Review, 2012,26 (3):65-68.

[3] GUO Y F,QUAN W M,WANG W Y,et al. Crack Diagnosis Method of Wind Turbine Blade Based on Convolution Neural Network with 3D Vibration Information Fusion. Journal of Optics,2020,40(22):140-148.

[4]MURT AGH P JBASU B BRODERICK B M. Mode acceleration approach for rotating wind turbine blades. Proc Inst of Mech Engrs Part K: J Multibody Dynamics 2004 218(3): 159-167.

[5]Junkai Guo, Jiaqi Zhang,Jianwen Wang, et al. Analysis of influence of centrifugal Load on Modal Characteristics of small wind turbine blade. Acta Energiae Solaris Sinica,2019,40(08):2148-2153.

[6]Lijun Shan, Yuan Liu. 3 mw wind turbine blade aerodynamic characteristics. Journal of dalian jiaotong university, 2017,38 (01):71-75.

[7]Jundong Huang, Hongjian Xia,Deyuan Li, et al. Analysis of the influence of structural mode and flutter characteristics of wind turbine blade sweepback. Journal of Solar Energy, 2021,42 (11): 273-279.

[8]Guoyu Hu, Wenlei Sun, Ping Dong. Analysis on Mechanical Characteristics of Composite Blade Structure of 5MW Wind Turbine Generator. Renewable Energy,2015,33 (06): 871-875.

[9]Xu C, Amano R S,LEE E K. Investigation of an axial fan-blade stress and vibration due to aerodynamic pres- sure field and centrifugal effects. Fluids and Thermal Engineering, 2004, 47(1):75—90.

[10]RAO A R,DUTTA B K. Blade vibration triggered by low load and high back pressure. Engineering Failure Analysis,2014,46(11) : 40-48.

[11]RAO A R, DUTTA B K. Non-intrusive method of detecting turbine blade vibration in an operating power plant// Proceedings of ISMA. Sydney, Australia: [s.n.] ,2010: 2937-2947.

[12]Chen Y,Mendoza A S E, Griffith D T. Experimental and numerical study of high-order complex curvature mode shape and mode coupling on a three-bladed wind turbine assembly. Mechanical Systems and Signal Processing,2021,160:107873.

[13]Xiangming Jin, Deping Gao, Xianxin Cai, et al. Vibration Reliability Analysis of Monolithic Centrifugal Impeller Blade. Journal of Aerospace Power, 2004, 19(5):610-613.

[14]Xueshi Yao. Research on dynamics of wind turbine blades based on rotational softening. Journal of Power Engineering, 2011,31 (03): 209-213.

[15]Junping Zhang. Research on Blade Damage Identification of Rotating Wind Turbine Based on Vibration. Harbin Institute of Technology, 2012.

Effect of Tip Clearance Size on the Cavitation Flow and Noise Characteristics of the NACA0009 Hydrofoil

Yudi Zhou

College of Mechanical and Electronic Engineering, Shandong University of Science and Technology, Qingdao 266590, China.

[1]yudizhou97@163.com.

Abstract: The noise characteristics of NACA0009 hydrofoil with tip clearance are discussed. Select SST k- ω Turbulence model and Schnerr Sauer cavitation model simulate tip leakage cavitation flow. The results show that the simulation results are in good agreement with the original records, and the tip clearance reduces the energy characteristics of the hydrofoil. With the expansion of tip clearance model, TLV and cavitation compressive strength gradually decrease. In addition, the study of far-field noise shows that the nonspecific curve of the total noise coefficient of far-field noise shows a disc shaped symmetric structure. Compared with the original hydrofoil, when the tip clearance is 20 mm, the far-field dipole noise value at 170 ° angle is reduced by 17.68%. Tip clearance reduces offline and far-field noise. Far field noise decreases with the increase of tip clearance model.

1. Introduction

In the naval ship, submarine and water conservancy and hydropower engineering industry [1], TLV cavitation and the noise caused by it, especially the cavitation and vibration of hydraulic machinery [2], seriously endanger the safe operation of hydraulic machinery, and have received more and more attention and concern. When cavitation occurs, cavitation noise will significantly reduce the hiding characteristics of underwater targets, and even become an extremely important noise source for the overall underwater targets [3]. Therefore, the study of TLV, cavitation and noise plays a vital role.

The research staff conducted many tests to reduce the adverse effects of TLV on water transmission equipment. They can also carry out research by accurately measuring the total tip leakage flow of the air compressor [4]. The results show that the strength of TLV changes with the size of tip clearance [5]. At the level of simulation calculation, scholars specifically applied Schnerr Sauer cavitation solid model and Rayleigh Plesset cavitation solid model to carry out simulation research, in order to test the evolution of TLV cavitation flow in detail, and analyze the impact of TLV trajectory and hydrofoil structure on TLV [6]. However, scholars mainly considered the accuracy of numerical calculation methods and the mechanism of TLV cavitation, while ignoring the interaction between tip clearance specifications and TLV cavitation and noise. In order to ensure the accuracy of noise prediction, scholars have taken into account the impact of turbulence models on prediction and analysis noise [7]. The research shows that the turbulence model harms the noise sources of single-stage and induced effects. Although scholars have studied a large number of factors affecting the prediction accuracy of noise, few scholars have fully considered the impact of tip clearance specifications on hydrofoil induced far field noise.

On the premise of previous research, this paper aims to study the flow regularity of NACA0009 hydrofoil among TLV, cavitation and noise after having tip

Raines. Using SST k- ω Turbulence model, Schnerr Sauer cavitation solid model and FW-H acoustic model are used to explain the mechanism of tip clearance specifications affecting TLV cavitation and noise, so as to better understand TLV, cavitation flow and noise.

2. Geometric Model and Mesh

2.1. Computational Fluid Domain Design

According to Dreyer's hydrofoil experiment [8], the length of drain l is 750 mm, the width s is 150 mm, and the height h is 150 mm. NACA0009 hydrofoil is used in the simulation experiment in this paper. The chord length c is 100 mm, and the thickness of the larger hydrofoil is 9.9 mm. Inclination angle of hydrofoil α 10 °, and the exit velocity U is 10 m/s. In the experiment, a gap was noted between the side port and the wall of the wing, with a width of w. The gap width is compared with the main thickness of the wing to obtain a dimensionless gap width $\tau = \delta/w$ In the customized experiment of this article, τ The values of are 0, 0.4, 1 and 2, and the corresponding configurations are M0, M4, M10 and M20. basis τ Numerical value, set up the difference calculation domain of simulation calculation. As shown in Figure 1 below, the top of the head with the chord length of the hydrofoil is located at z=0, and the positions of different liquidity orientations can be determined by the chord length ratio z/c in the z coordinate domain.

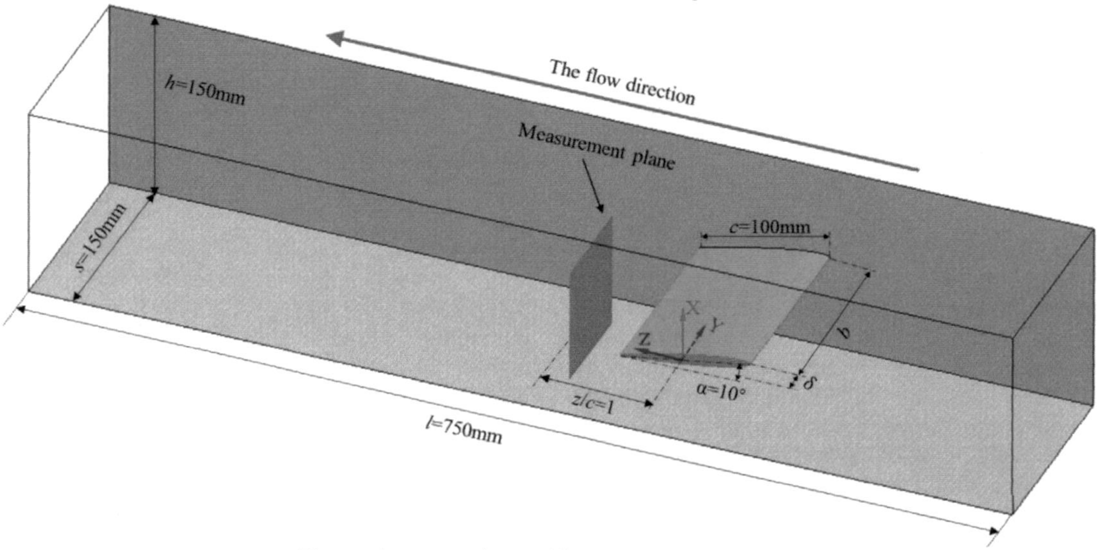

Figure 1. Stereoview with measurement plane.

2.2. Numerical Setting

STAR CCM mobile phone software is selected to carry out numerical calculation method. In order to capture the small details of TLV fluidity, the turbulence model is selected as SST k- ω Solid model and cavitation solid model are selected as Schnerr Sauer model. According to the previous experimental environment [8], the initial inlet condition is set as the velocity inlet, and the inlet velocity is 10 m/s. The initial outlet condition is set to hydrostatic outlet. The hydrofoil surface and swallow hole wall are set as ground anti-skid boundary layer. In the solution phase, the iterative updating process of the larger internal structure is set as 20 stages.

2.3. Mesh Independence Test

According to the number of grids commonly used in reference [1], 5.0 grids are selected for leaf clearance × 106 Grid. Because it is difficult to obtain cavitation experimental conclusions for NACA0009 hydrofoil without leaf gap, this paper conducts grid incoherence research on NACA0009 hydrofoil without leaf gap to ensure the accuracy of numerical value. The lift and friction resistance of

NACA0009 hydrofoil with different meshes are shown in Table 1. When the number of grids exceeds 4.7 × At 106, the lift error is kept within 0.1%, and the friction resistance error is kept within 1%. This indicates that when the number of grids exceeds 4.7 × At 106, the calculated precision is very high. Fully consider the accuracy of the calculation cloud computing server, and select a total grid value of 4.7 × one hundred and six.

Table 1. Lift and drag under different mesh numbers.

Mesh scheme	Mesh1	Mesh2	Mesh3	Mesh4	Mesh5
Mesh number	1.5×10^6	2.7×10^6	4.7×10^6	6.7×10^6	8.7×10^6
Lift	943.902	954.297	954.713	954.851	954.925
Change Lift (%)	−1.154	−0.066	−0.022	−0.008	–
Drag	24.062	22.92	22.965	22.944	22.811
Change Drag (%)	5.484	0.478	0.625	0.583	–

3. Results and Discussion

3.1. Validation of Numerical Method

In order to verify the correctness of the numerical calculation process of hydrofoil cavitation flow, the numerical calculation process and results of the hydrofoil in this paper are compared with the leaf gap specification. As shown in Figure 2 below, when the inlet velocity is 10m/s and the tilt angle is 10 °, the TLV cavitation obtained from the experiment and the TLV cavitation obtained from the simulation calculation (based on the vapor mole fraction α V is defined as 0.1). The numerical calculation clearly predicted and analyzed the TLV position.

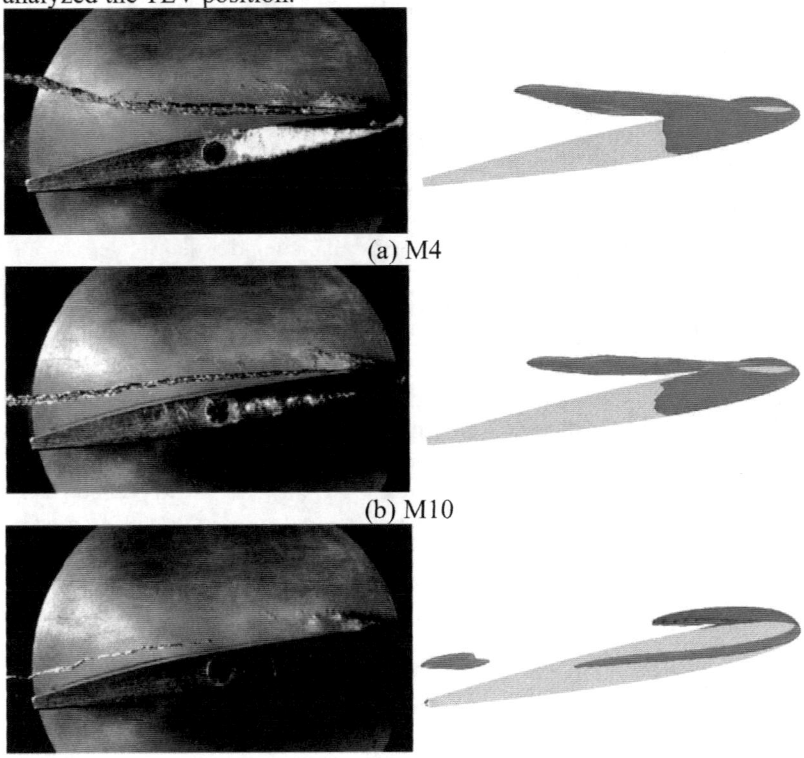

(a) M4

(b) M10

(c) M20

Figure 2. Comparison of cavitation vortex between experiment and simulation results.

3.2. Comparison of energy characteristics of hydrofoils with different tip clearances

The results show that the cavitation potential flow of hydrofoil is a cavitation potential flow. Input speed 10m/s, dip Angle 10°. Numerical calculations clearly predict, analyze, and describe magnetic positions:

$$C_L = \frac{L}{0.5\rho U^2 S_L} \tag{1}$$

$$C_D = \frac{D}{0.5\rho U^2 S_D} \tag{2}$$

$$K = \frac{C_L}{C_D} \tag{3}$$

In which, L and D respectively indicate the friction resistance and lift force perpendicular to the inflow direction and perpendicular to the inflow direction. They are the forces in the X axis and Z axis directions in the current simulation. SL and SD respectively indicate the developed area of the hydrofoil in the X and Z directions.

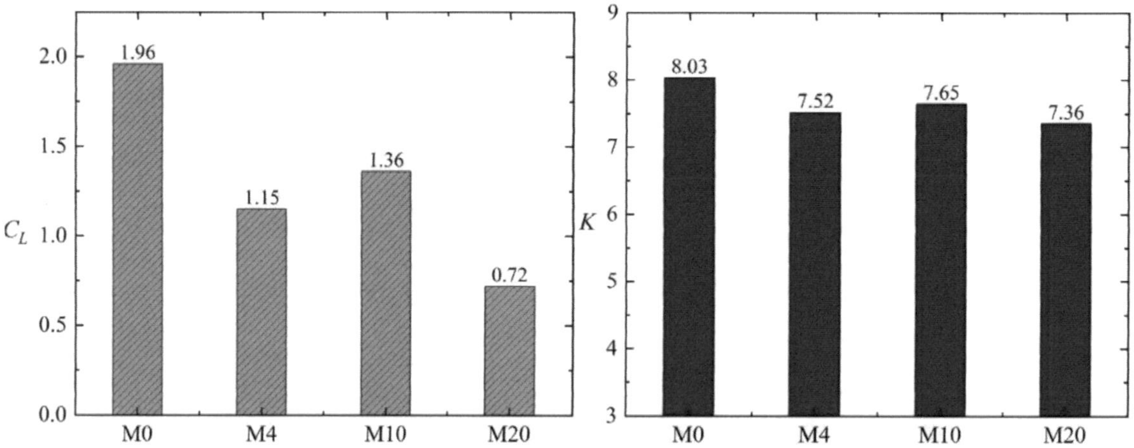

(a) Lift coefficient (b) Ratio K of lift coefficient to drag coefficient

Figure 3. Variation of hydrofoil energy performance under the tip clearance size.

3.3. Comparison of cavitation characteristics of hydrofoils with different tip clearances

In order to scientifically study the influence of tip clearance specification on cavitation flow characteristics, the larger cavitation torque in the second cavitation falling circulation system in Figure 4 is selected for comparison. Figure 4 shows the tip clearance specification (vapor mole fraction) when large cavitation occurs α V=0.1 equivalent circuit surface). The cavitation of the surface covers the relatively local characteristics of the hydrofoil surface. In the area near the tip clearance, the cavitation refractive index with bending characteristics increases. This shows that with the increase of tip clearance specification, the cavitation level on the surface area of the hydrofoil gradually decreases. Figure 4 also shows that with the expansion of tip clearance specification, the TLV cavitation intensity tends to increase first and then decrease. Therefore, the change of tip clearance may affect the intensity of TLV cavitation.

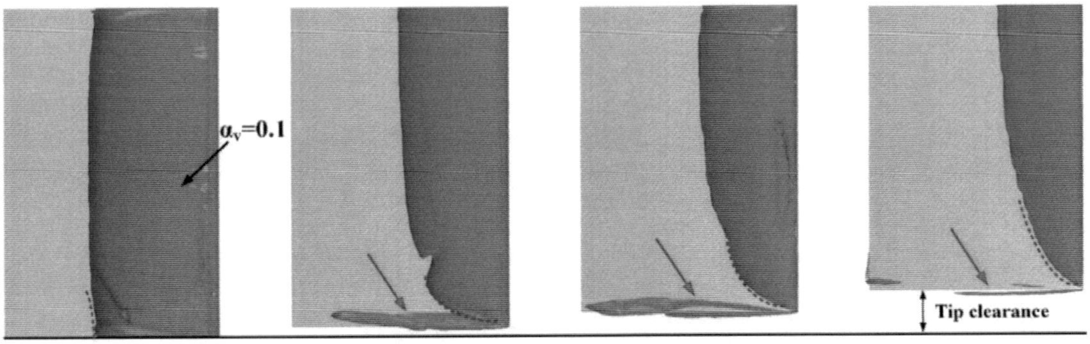

(a) M0 (b) M4 (c) M10 (d) M20

Figure 4. Comparison of sheet cavities from a top view.

3.4. Effect of Tip Clearance Size on far-field radiation noise

In order to scientifically study the influence of tip clearance specification on the far-field noise characteristics, the larger cavitation torque in the second cavitation falling circulation system in Figure 5 is selected for comparison. In order to energetically analyze the specificity of far field hydrofoil radiation source noise under tip clearance specification, 72 monitoring points are selected in this paper, and the overall noise coefficient (OASPL) of each monitoring point is described. As shown in Figure 5 below, the observation stations are jointly distributed in a circle with a half diameter of 1m.

OASPL can be expressed as:

$$\mathrm{OASPL} = 20\log\left(\frac{P_{rms}}{P_{ref}}\right) \tag{4}$$

where P_{ref}=1.1427×10^{-6} Pa is the reference sound pressure, and P_{rms} is the effective sound pressure in Pa.

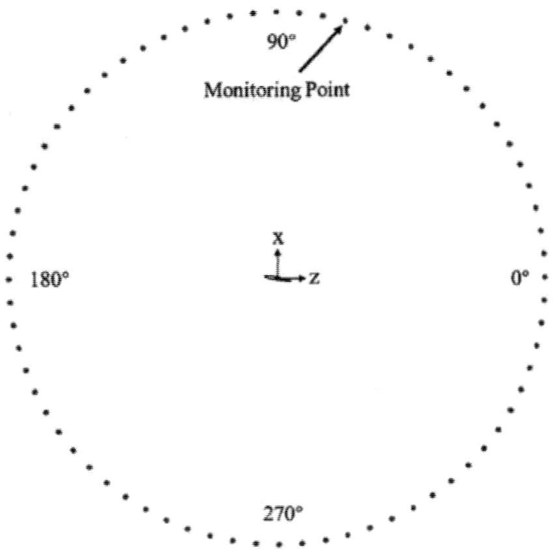

Figure 5. Acoustic field points on the plane x-z.

Figure 6 shows OASPL specificity curves of 72 monitoring points in case of large cavitation erosion. The guidance curve of OASPL shows that symmetrical butterflies are everywhere. OASPL is also larger at 80 ° and 260 ° of the hydrofoil vertically, and smaller at 170 ° and 350 °. The fundamental reason for this is that the dipole noise source is mainly distributed on the adsorption surface and working pressure surface of the hydrofoil. Figure 6 also shows a large OASPL value in the initial hydrofoil. With the increase of tip clearance specification, the dipole noise at the far field monitoring point tends to be smaller and smaller, indicating that the increase of tip clearance specification has a certain sound insulation effect.

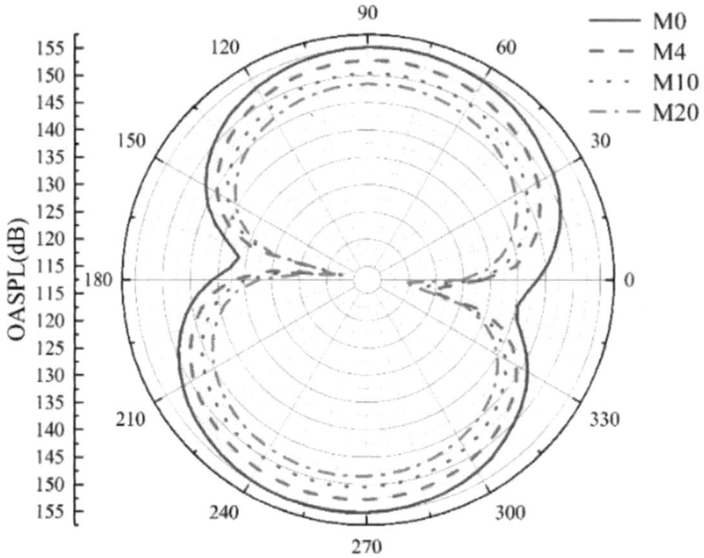

Figure 6. Directionality of OASPL under the tip clearance size.

Table 2 shows the impact of different leaf gap specifications on OASAP of hydrofoils. The relative transformation is defined as

Δ OASPL= (OASAP-OASAP0) /OASAP0 (5)

OASAP0 is the total sound intensity under M0.

As shown in Table 2, compared with the initial hydrofoil, the far-field dipole noise at the angles of 80 °, 170 °, 260 ° and 350 ° at M20 decreases by more than 10%. In particular, the far-field dipole noise at 170 ° is reduced by 17.68%. Compared with the initial hydrofoil, the far field induced noise value of M4 decreases by less than 10% in the directions of 80 °, 170 °, 260 ° and 350 °, and the far field bipolar noise value of M10 decreases by about 10%. Compared with the initial hydrofoil, under the conditions of M4, M10 and M20, the far-field dipole noise at 80 °, 170 °, 260 ° and 350 ° shows a downward trend of dipole noise. This finding shows that the tip gap significantly reduces the far-field dipole noise. In addition, the far field dipole noise value gradually decreases with the expansion of the leaf gap specification.

Table 2. OASPL for tip clearance size.

ΔOASPL%	M4	M10	M20
0°	–6.98	–7.5	–9.61
170°	–12.62	–14.25	–13.22
350°	–11.59	–12.87	–12.65
355°	–10.42	–12.33	–15.93

4. Conclusions

This paper discusses the harm of different tip clearance dimensions to the sound field characteristics of NACA0009 hydrofoil energy performance. Through the verification of the numerical calculation method, the predicted TLV cavitation erosion is in good agreement with the experimental observation conclusion. The specific results are as follows:

(1) With the increase of tip clearance size, the lift coefficient and lift drag ratio show a significant downward trend. Compared with the initial hydrofoil, the lift coefficients under M4, M10 and M20

conditions decreased by 41.1%, 30.6% and 63.1% respectively, indicating that the tip clearance size significantly harmed the energy characteristics of the hydrofoil.

(2) With the increase of tip clearance size, TLV strength tends to increase first and then decrease. In addition, along with the increase of tip clearance size, the cavitation erosion of the hydrofoil area slowly decreases, and the TLV cavitation has a trend of first increasing and then decreasing.

(3) For far-field noise, OASPL's specificity is distributed with symmetrical butterfly patterns. Compared with the initial hydrofoil, under the premise of M20, the far field dipole noise value at 170 ° angle is reduced by 17.68%. The tip gap reduces the decibel number of far-field dipole noise. With the increase of tip clearance size, the far-field dipole noise tends to be smaller and smaller, indicating that increasing tip clearance size has a certain sound insulation effect.

References

[1] Chen Y, Zhang W, Fang D, et al. Vortex suppression and flow pattern analysis of a hydrofoil with parallel grooves[J]. Processes, 2021, 9(5): 816.

[2] Tao R, Xiao R, Wang F, et al. Cavitation behavior study in the pump mode of a reversible pump-turbine[J]. Renewable Energy, 2018, 655–667.

[3] Lu J, Liu X, Zeng Y, et al. Investigation of the noise induced by unstable flow in a centrifugal pump[J]. Energies, 2020, 13(3): 589.

[4] Inoue M, Kuroumaru M, Fukuhara M, et al. Behavior of tip leakage flow behind an axial compressor rotor[J]. Journal of Engineering for Gas, 1986, 108(1): 7–14.

[5] Storer J, Cumpsty N. Tip leakage flow in axial compressors[J]. Journal of Turbomachinery, 1991, 113(2): 252–259.

[6] Cheng H, Long X, Ji B, et al. Suppressing tip-leakage vortex cavitation by overhanging grooves[J]. Experiments in Fluids, 2020, 61: 159.

[7] Yu A, Wang Z, Zou Z, et al. Investigation of cavitation noise in cavitating flows around a NACA0015 hydrofoil[J]. Applied Sciences, 2019, 9(18): 3736.

[8] Dreyer M, Decaix J, Mohamed F, et al. Mind the gap: a new insight into the tip leakage vortex using stereo-PIV[J]. Experiments in Fluids, 2014, 55: 1849.

Journal of Physics: Conference Series **2455** (2023) 012014 doi:10.1088/1742-6596/2455/1/012014

Calculation and analysis of fatigue strength of truss-floating box aquaculture vessels

Yuchen Zhang[1,a], Zhaode Zhang[1]

[1]School of Ship and Maritime, Zhejiang Ocean University Zhoushan, China

[a]e-mail: 793996831@qq.com

Abstract. In this paper, DNVGL's SESAM software system is used to directly calculate the fatigue strength of typical structural nodes of truss-floating box aquaculture vessels based on fatigue spectrum analysis. According to the three-dimensional potential flow theory, the hydrodynamic analysis of typical working conditions of the ship type was carried out, and the short-term distribution of stress response transfer function, stress response spectrum and stress range under culture conditions was obtained. The load is applied to the ship-wide finite element model to obtain the fatigue strength stress values of typical thermal stress points, and then the fatigue life is estimated by spectral analysis. The research results can provide useful reference and reference for the design and verification stage of the ship type.

1. Introduction

Chao Ma[1] used potential flow theory and dynamic composite cable theory to simulate the hydrodynamic responses of a single-point moored vessel-shaped floating aquaculture platform in waves; The structural hot-spot stress as well as the effective notch stress approach have been applied by Wolfgang Frick[2]. Insofar the investigations give a good insight into the strength behaviour of complex welded structures and into current problems and opportunities offered by numerical analyses; Based on test and finite element method, Myung Hyun Kim[3] studied the fatigue strength of long web joint welds in detail, hoping to develop more accurate fatigue strength evaluation techniques; A. Yosri[4] investigated the fatigue life of structural details in the double side of a tanker ship, employing the spectral fatigue analysis approach. The influences of different wave scatter diagrams, reduction of hull girder section modulus, stress concentration factor and free corrosion S–N curve on the accumulated fatigue damage were investigated.

The concept of developing large-scale marine fishery aquaculture equipment was proposed in Western countries in the 80s of the last century, mainly including floating platforms, shipborne aquaculture workshops, cabin aquaculture and semi-submersible cage platforms. Nowadays, large-scale aquaculture vessels mainly have two forms: swimming shipborne aquaculture platform and cage aquaculture platform, and these two types of aquaculture vessels have their own characteristics and advantages and disadvantages. In order to better meet the two working conditions of towing and aquaculture, on the basis of the current two types of aquaculture vessels, there is a new type of truss-floating box aquaculture vessel mentioned in the text.

In this paper, the stress level and stress distribution of the local components of the truss-floating tank aquaculture vessel under different sea conditions are calculated by the finite element method, and the calculation results are more accurate and in line with the actual navigation, and then the analysis and evaluation of the fatigue strength of the hull are reasonably completed based on the fatigue spectrum analysis method.

2. Finite element model building

2.1. Finite element model master scale and calculation conditions

Considering that most of the use environment of aquaculture vessels is cultured, the aquaculture conditions are calculated as typical working conditions in this paper. Establish a three-dimensional wet surface model of the hull, Morrison model and mass model according to the working condition parameters. The draft depth is simulated with the wet surface model, and the displacement is simulated with the quality bar model. The size of the target ship owner is shown in Table 1, and the draft parameters in typical working conditions are shown in Table 2.

Table 1. Target owner scale.

Primary metrics	Value	Primary metrics	Value
Captain/m	129.00	Molded wide/m	40.00
Deep/m	11.20	Design draft/m	10.00

Table 2. Calculate the operating condition parameters.

Working conditions	Displacement/(t)	Bow draft/(m)	Stern draft/(m)
Breeding conditions	2819.6	10.003	10.017

2.2. Finite element model and fatigue check points

According to the relevant requirements of finite element modeling in CCS "Guidelines for the Fatigue Strength of Hull Structures", the whole ship finite element model of the ship type was established by DNV.SESAM software. The finite element model of the whole ship structure includes the bow flotation box structure, buttressed timber, strong frame and amidship truss structure. The finite element model of the coarse mesh structure has a total of 49189 elements and 31509 nodes, which are shown in Figure 1(a).

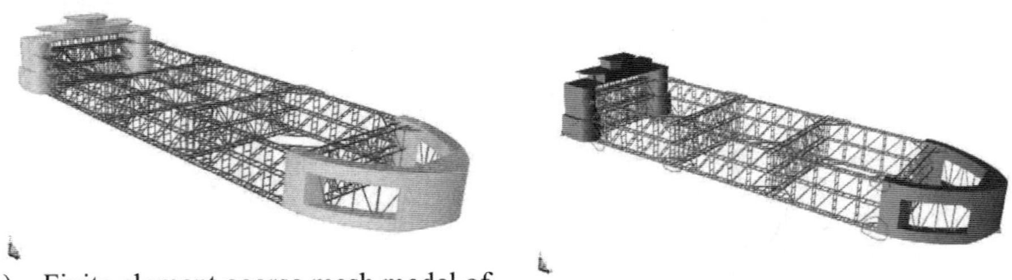

(a) Finite element coarse mesh model of the whole ship (b) Fatigue check point location

Figure 1. Finite element model of the whole ship.

Due to the symmetry of the hull, 2 check points in each of the 6 positions were selected, a total of 12 typical check points, and the fatigue check point positions are shown in Figure 2 (b). 2 check points (numbers 1 and 2) at the connection between the tail truss and the upper part of the floating tank; 2 check points (numbers 3 and 4) at the junction of the tail truss and the lower part of the floating tank; 2 check points (numbers 5 and 6) at the junction of the longitudinal truss and the upper part of the transverse frame; 2 check points (Nos. 7 and 8) at the lower junction of the longitudinal truss and the transverse frame; 2 check points (Nos. 9 and 10) at the connection between the head truss and the upper part of the floating tank; 2 check points at the junction of the head truss and the lower part of the floating tank (Nos. 11 and 12).

2.3. Refine the finite element model

Refine the grid principle: (1) The size of the grid cannot exceed the thickness value of the stressed member; (2) The scope of the mesh refinement area should ensure that the extension from the fatigue hot spot position to all directions is not less than 10 times the thickness of the component plate; (3) The panels and webs of beam units near the hot spot should be simulated by plate and shell elements; (4) Use 4-node shell elements at hot spots to avoid the use of triangular elements; (5) A 1:1 element aspect ratio is used in hot spots and areas very close to hot spots, the aspect ratio of units in areas slightly away from hot spots is limited to 1:3, and the inner angle of four gusset elements or shell elements is limited to between 50 degrees and 130 degrees. The refinement model of each check point is shown in Figure 2.

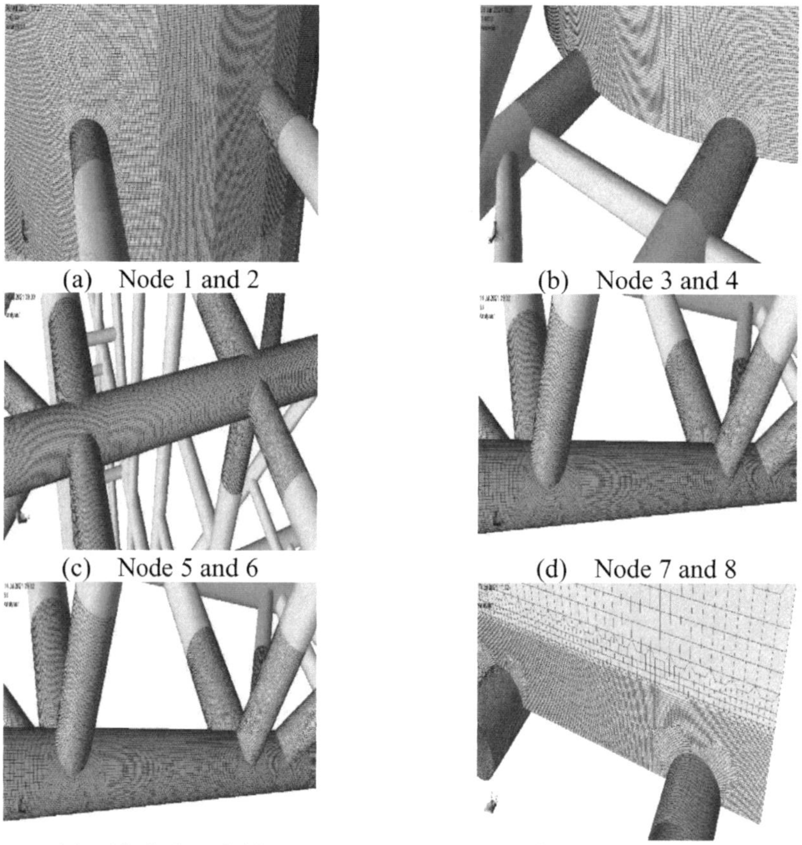

(a) Node 1 and 2 (b) Node 3 and 4

(c) Node 5 and 6 (d) Node 7 and 8

(e) Node 9 and 10 (f) Node 11 and 12

Figure 2. Check points to refine the mesh model.

3. Fatigue spectroscopy analysis

3.1. Principles of spectral analysis

The direct calculation method based on spectral analysis is a commonly used method to study the load and structural response in ship and offshore engineering, and the spectral analysis method first calculates the wave load and the stress response of the hull structure under different loading conditions, wave direction and calculation frequency of the target ship, and then obtains the stress transfer function. Then, combined with the wave spectrum and sea state data of specific sea areas, the stress response spectrum and the short-term distribution of stress are obtained. Finally, based on the linear cumulative damage theory and the corresponding S-N curve, the total cumulative damage degree of the fatigue evaluation site is calculated. The flow of spectroscopy using SESAM is shown in Figure 3

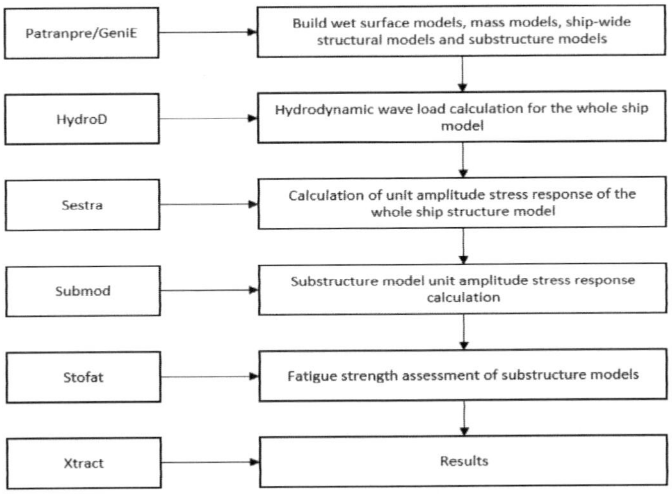

Figure 3. Fatigue assessment flowchart using SESAM.

3.2. Wave load response calculation

When calculating the speed of the breeding condition, it is 0kn. The wave direction angle is 0°~180°, and the interval is 30°, with a total of 12 wave directions. The specific calculation conditions and parameters of the determined wave load response of the ship are shown in Table 3.

Table 3. Wave load response parameters.

project	unit	content
Speed	Knots	0
Wave angle	Deg	0°、30°、60°、90°、120°、150°、180°
Probability of occurrence	-	1/7
Wave cycle	s	1、2、3、4、5、6、7、8、9、10
Wavy spectrum	-	P-M spectrum

4. Calculation results

4.1. Wave load response calculation

The calculated fatigue load is applied to the finite element model of the hull structure, and the stress analysis of the finite element model of the hull is carried out using the inertia release function of SESAM software to obtain the stress response transfer function of the hull structure under each wave spectrum.

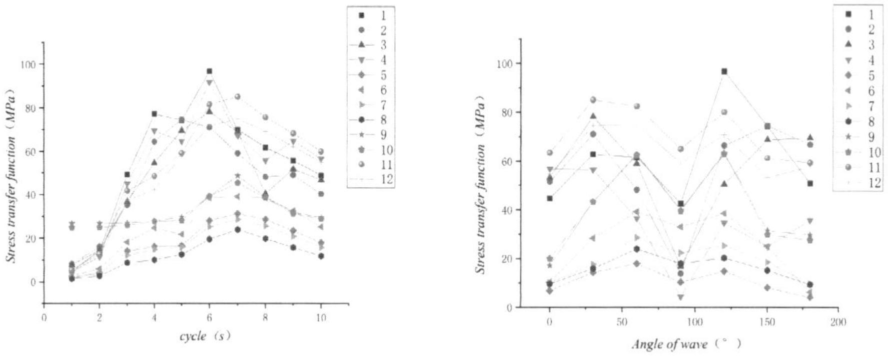

(a) The stress transfer function varies with cycles (b) The stress transfer function varies with the wave direction

Figure 4. The result of the transfer function of the stress response of each hot spot.

The stress transfer function of the analyzed junction varies with the period as shown in Figure 4(a). It can be seen from the figure that the stress response transfer function of each node in the period of 1-7s increases with the increase of the period, and the stress response transfer function obtains the maximum value when T=7s. It decreases after 7 seconds of the cycle. Among the 1-12 nodes, node 1 has the largest stress response transfer function of 96.722MPa, and the wave direction angle at this time is 120°. But the stress response transfer function of node 1 plummets down after 7s. Therefore, this ship type is the worst situation of this ship type when it encounters waves with a period of about 7s.

The stress transfer function of the analyzed junction varies with the wave direction angle as shown in Figure 4(b). It can be seen from the figure that the maximum stress transfer function value of 12 nodes at the same wave direction angle is not much different and the trend changes with the wave direction angle is similar. Except for node 1 and node 2, the maximum value of all other nodes occurs at 30° wave direction. The maximum value of node 1 occurs at 120° and the maximum value of node 2 occurs at 150°. And the minimum value of the 12 nodes all occurs at the wave direction angle of 90°.

4.2. Degree of fatigue damage

When evaluating the fatigue strength of the ship structure by spectral analysis, the stress response spectrum can be obtained by obtaining the amplitude-frequency response function and combining the power spectral density function of the wave. The power spectral density function of a wave generally uses the biparametric spectrum recommended by the International Conference on Ship Structures (ISSC). In this paper, the wave dispersion map of global sea state is used to analyze the fatigue of the ship structure, and the fatigue cumulative damage degree at each node is calculated by combining Miner's linear cumulative damage theory, as shown in Figure 5. The cumulative fatigue damage degree D of the recovery period is calculated from the obtained fatigue, and the actual fatigue life can be obtained.

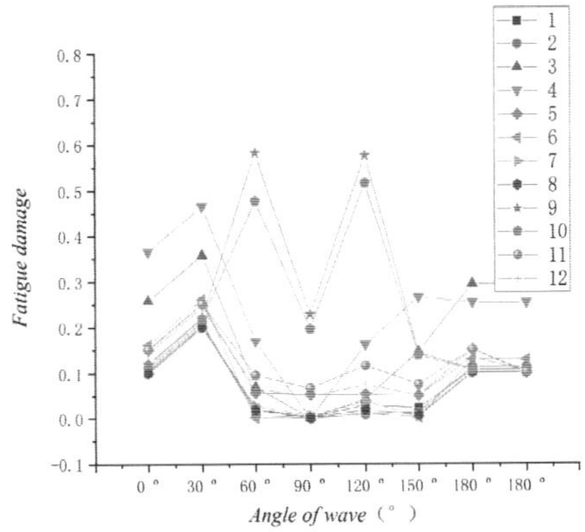

Figure 5. The degree of fatigue damage varies with the wave direction.

It can be seen from the figure that the trend of nodes 1, 2, 5, 6, 7, 8, 9, 11, and 12 with the wave direction angle is small and the values are similar. The minimum values of these 8 nodes appear at 0° (following the wave) and 180° (against the wave), the values are around 0.01; the maximum value occurs at 120°, the values are around 0.05. Nodes 3, 4, 9, and 10 have a large trend with the wave direction angle and the change trend is not synchronized. Nodes 9 and 10 have the same change trend, and the fatigue damage degree will obtain the maximum and minimum values at the same time. The maximum value occurs at 60° and 120°, and the minimum occurs at 90°. Node 3 and node 4 maximum occurs at 30° and minimum at 90°. In summary, 12 nodes are at a wave angle of 90° at the same time, and the fatigue damage degree is minimal, so when performing breeding tasks for a long time, facing the waves with an attitude of 90 will extend the fatigue life of the ship.

4.3. Fatigue life calculation

The truss-floating box aquaculture vessel has a design life of 25 years, and the fatigue life is the ratio of the design life to the fatigue cumulative damage, and the calculation results are shown in Table 4.

Table 4. Fatigue life.

location	Node	Fatigue life（year）
The tail truss is connected to the upper part of the floating tank	1	40.3
	2	48.0
The junction of the tail truss with the lower part of the floating tank	3	43.0
	4	29.5
The longitudinal truss joins the upper part of the transverse frame	5	32.7
	6	42.5
The longitudinal truss joins the lower part of the transverse frame	7	45.6
	8	65.7
The junction of the head truss with the upper part of the floating tank	9	28.9
	10	31.4
The junction of the head truss with the lower part of the floating tank	11	29.0
	12	32.1

From the results, it can be seen that the fatigue life of the 12 nodes is greater than the design life, that is, the fatigue strength of the truss-floating box aquaculture vessel meets the requirements of the specification.

5. Conclusion

Because the aquaculture vessel is different from the traditional shipborne breeding platform and cage breeding flat, it is a new truss-floating box ship structure. In order to ensure that the aquaculture vessel has sufficient fatigue strength safety margin in the breeding process, it is necessary to directly calculate

Journal of Physics: Conference Series **2455** (2023) 012014 doi:10.1088/1742-6596/2455/1/012014

the fatigue strength of the ship's hazard profile in the design. In this paper, the fatigue strength of the central truss is evaluated by direct calculation method based on spectral analysis, and the calculation results show that: 1) the fatigue life of the thermal stress point can meet the requirements of the design service life and has a certain safety margin; 2) The life of the connection between the longitudinal bone and the floating box is short; 3) The fatigue life of the underwater longitudinal bone is shorter than that of the longitudinal bone above the waterline surface; 4) Therefore, these are the positions that should be considered when the fatigue strength check of this ship type is carried out in the future.

REFERENCES

[1] Chao Ma and Yun-Peng Zhao and Chun-Wei Bi 2020 Numerical study on hydrodynamic responses of a single-point moored vessel-shaped floating aquaculture platform in waves J. Aquacultural Engineering, (2022)096

[2] Wolfgang Frick and Hans Paetzold 2010 Full-scale fatigue tests of ship structures to validate the S–N approaches for fatigue strength assessment J Marine Structures, JHEP01(2010)023

[3] Myung Hyun Kim and Sung Won Kang and Jeong Hwan Kim and Kwang Seok Kim and Joong Kyoo Kang and Joo Ho Heo 2010 An experimental study on the fatigue strength assessment of longi-web connections in ship structures using structural stress J International Journal of Fatigue, JHEP02(2010)032

[4] A. Yosri and H. Leheta and S. Saad-Eldeen and A. Zayed 2022 Accumulated fatigue damage assessment of side structural details in a double hull tanker based on spectral fatigue analysis approach J Ocean Engineering, (2022)251

Journal of Physics: Conference Series **2455** (2023) 012015 doi:10.1088/1742-6596/2455/1/012015

Estimates for the minimum eigenvalue of an M-matrix

Qin Zhong [1] **and Ling Li** [2*]

[1] Department of Mathematics, Sichuan University Jinjiang College, Meishan 620860, China

[2] School of Big Data and Artificial Intelligence, Chengdu Technological University, Chengdu 611730, China

[*] Corresponding author's e-mail: lynn910309@126.com

Abstract: New sharper lower bounds for the minimal eigenvalues of symmetric M-matrices and nonsymmetric M-matrices are proposed by constructing two increasing sequences. By contrasting the proposed instances with the related findings, two instances are offered to demonstrate the effectiveness of the technique.

1. Introduction

Let $N=\{1,2,\cdots,n\}$, the matrix A with the form

$$A=\alpha I-P, \tag{1}$$

in which α is a real number and P a nonnegative matrix, is defined as a Z-matrix. Let $\sigma(P)$ be the spectrum of the matrix P and

$$\rho(P)=\max\{\mathrm{Re}(\lambda):\lambda\in\sigma(P)\}.$$

The matrix A is referred to as an invertible M-matrix when $\alpha>\rho(P)$ [1]. The collection of M-matrices of order n is denoted by M_n. The following expression is

$$q(A)=\min\{\mathrm{Re}(\lambda):\lambda\in\sigma(A)\}.$$

is called the minimal eigenvalue of the matrix A. Furthermore, in reference [1], it is conducted that if $A\in M_n$, then $q(A)\in\sigma(A)$.

Many experts and scholars have carried out extensive research on M-matrices [2-5]. A beautiful conclusion is that if $A\in M_n$ and if it is defined in (1), then

$$q(A)=\alpha-\rho(P).$$

The goal of this study is to suggest some lower bounds on $q(A)$ that are more precisely defined for an M-matrix A. The structure of the article is as follows. The introduction is the first section. In Section 2, for an M-matrix, we investigate the lower bounds for the minimal eigenvalue. Moreover, two increasing sequences of lower bounds for $q(A)$ are presented in Section 3. In Section 4, two instances are added to the comparison of lower bounds with known ones.

2. Lower bounds for the minimum eigenvalue of an M-matrix

We first introduce the following well-known results from the reference [4].

Theorem 2.1 If the matrix $A=(a_{ij})$ is weakly chained diagonally dominant and $A^{-1}=(\beta_{ij})$, then

(1) $q(A)\geq \min\left\{\sum_{j\in N} a_{ij} : i\in N\right\}$,

(2) $\frac{1}{M}\leq q(A)\leq \frac{1}{m}$,

where $M=\max_{i\in N}\sum_{j\in N}\beta_{ij}=\left\|A^{-1}\right\|_{\infty}$ and $m=\min_{i\in N}\sum_{j\in N}\beta_{ij}$.

To estimate the bound for $q(A)$, in [4], the following important result has been conducted: if the n-square matrix A is an M-matrix that satisfies the condition of weakly chained diagonally dominant, then

$$\left\|A^{-1}\right\|_{\infty}\leq \frac{1}{a_{11}(1-r_1)}+\frac{1}{a_{22}(1-r_1)(1-r_2)}+\cdots+\frac{1}{a_{nn}(1-r_1)\cdots(1-r_n)}, \tag{2}$$

where $r_k=\frac{1}{|a_{kk}|}\sum_{j=k+1}^{n}|a_{kj}|<1, k\in N$ and $r_n=0$.

The following result is due to reference [5]:

Theorem 2.2 Let $A=(a_{ij})$ be an irreducible M-matrix and let $\alpha=\max_{i\in N} a_{ii}$, $R_i(A)=\sum_{j=1} a_{ij}$, and $M_i(A)=\sum_{j=1} a_{ij}R_j(A)$. Then

$$\min_{i\in N} h_i(A)\leq q(A)\leq \max_{i\in N} h_i(A),$$

where $h_i(A)=\frac{\alpha R_i(A)-M_i(A)}{\alpha-R_i(A)}$.

For the n-square matrix A with real eigenvalues $\lambda_1\geq\lambda_2\geq\cdots\geq\lambda_n$, in reference [6], it gives the following result:

$$\left|\lambda_i-\frac{trA}{n}\right|\leq\left[\frac{n-1}{n}\left(trA^2-\frac{(trA)^2}{n}\right)\right]^{1/2},$$

in which the symbol trA represents the trace of the matrix A. In particular,

$$\left|\lambda_n-\frac{trA}{n}\right|\leq\left[\frac{n-1}{n}\left(trA^2-\frac{(trA)^2}{n}\right)\right]^{1/2}. \tag{3}$$

As a result, we can quickly deduce the following conclusion from (3).

Theorem 2.3 Let the n-square matrix A be an M-matrix with real eigenvalues, then

$$q(A) \geq \frac{trA}{n} - \left[\frac{n-1}{n}\left(trA^2 - \frac{(trA)^2}{n}\right)\right]^{1/2}. \tag{4}$$

For the matrix A, in [7], the following notion of an arithmetic symmetrized matrix has been proposed:

$$M(A) = \frac{A + A^T}{2}.$$

It is further deduced that if A is a nonnegative matrix, then

$$\rho(A) \leq \rho\left(\frac{A + A^T}{2}\right).$$

Here we show that, for an M-matrix, we have a similar property of the minimal eigenvalue.

Theorem 2.4 Let $M(A) = \frac{A + A^T}{2}$ be the arithmetic symmetrization of the M-matrix A. Then

$$q(A) \geq q(M(A)). \tag{5}$$

Proof Let $A = \alpha I - P$, P is nonnegative, $\alpha > \rho(P)$. Then

$$M(A) = \alpha I - \frac{P + P^T}{2}.$$

Since $\alpha I - \frac{P + P^T}{2}$ is a Z-matrix, which has a real eigenvalue $q\left(\alpha I - \frac{P + P^T}{2}\right)$ with minimal real parts, we have

$$q\left(\alpha I - \frac{P + P^T}{2}\right) = \alpha - \rho\left(\frac{P + P^T}{2}\right).$$

Noting that $\rho(P) \leq \rho\left(\frac{P + P^T}{2}\right)$, thus we can derive

$$q\left(\alpha I - \frac{P + P^T}{2}\right) \leq \alpha - \rho(P).$$

Recalling the fact that $q(A) = \alpha - \rho(P)$, we achieve our desired conclusion.

Theorem 2.5 Let the matrix A be an M-matrix of order n. Then

$$q(A) \geq \frac{trA}{n} - \left[\frac{n-1}{n}\left(tr(M(A))^2 - \frac{(trA)^2}{n}\right)\right]^{1/2}, \tag{6}$$

where $M(A) = \frac{A + A^T}{2}$.

Proof Taking into consideration that the matrix $M(A)$ is symmetric and the symmetric matrix's eigenvalues are real numbers, based on the fact that $\frac{trA}{n} = \frac{trM(A)}{n}$, according to Theorem 2.3 and Theorem 2.4, Inequality (6) is immediate.

3. Increasing Sequences of Lower bounds for the Minimum eigenvalue of an M-matrix

In reference [8], the following important result has been proposed. All the eigenvalues of the complex matrix A of order n, whose eigenvalues are real numbers, lie in the intervals

$$\left[\frac{trA}{n}-\alpha_p, \frac{trA}{n}+\alpha_p\right], \tag{7}$$

where p is a natural number and

$$\alpha_p = \left[\frac{(n-1)^{2p-1}}{(n-1)^{2p-1}+1} tr\left(A-\frac{trA}{n}I\right)^{2p}\right]^{1/2p}. \tag{8}$$

Moreover, $\{\alpha_p\}$ is a decreasing sequence and

$$\alpha = \lim_{p\to\infty}\alpha_p = \max_i\left|\lambda_i - \frac{trA}{n}\right|, \tag{9}$$

in which λ_i is any eigenvalue of the matrix A. Now we give the main results of this part. For $A \in M_n$, firstly we consider the case when A has a real spectrum.

Theorem 3.1 If the n-square matrix A is an M-matrix with real eigenvalues, then for any natural number p,

$$q(A) \ge u_p \ge u_{p-1} \ge \cdots \ge u_2 \ge u_1, \tag{10}$$

where $u_p = \frac{trA}{n} - \alpha_p$ and

$$\alpha_p = \left[\frac{(n-1)^{2p-1}}{(n-1)^{2p-1}+1} tr\left(A-\frac{trA}{n}I\right)^{2p}\right]^{1/2p}.$$

Proof Since A has a real spectrum, in terms of (7), we have

$$\left|\lambda_j - \frac{trA}{n}\right| \le \alpha_p \tag{11}$$

for each eigenvalue λ_j of A and α_p is defined in (8). It is clear $q(A) \le \frac{trA}{n}$. Thereby, from Inequality (11) we acquire

$$q(A) \ge \frac{trA}{n} - \alpha_p.$$

Let

$$u_p = \frac{trA}{n} - \alpha_p.$$

Since the sequence $\{\alpha_p\}$ is monotonically decreasing, we have

$$q(A) \ge u_p \ge u_{p-1} \ge \cdots \ge u_2 \ge u_1.$$

This completes the proof.

Remark 3.1 If we set $p=1$ in (10) and note that

$$tr\left(A-\frac{trA}{n}I\right)^2 = tr\left(A^2\right) - \frac{(trA)^2}{n},$$

then Theorem 3.1 yields to Theorem 2.3.

Theorem 3.1 states that we may simply estimate the lower bounds of $q(A)$ if the matrix A is an M-matrix with real eigenvalues. One typical case is that A is a symmetric M-matrix, then we can use the method of Theorem 3.1 to estimate the minimum eigenvalue $q(A)$.

If the eigenvalues of the matrix A are not real, we just need to apply Theorem 3.1 to the arithmetic symmetrization of A. At this point, we will get the following results.

Theorem 3.2 Let the n-square matrix A be an M-matrix. Then for any natural number p,

$$q(A) \geq v_p \geq v_{p-1} \geq \cdots \geq v_2 \geq v_1, \tag{12}$$

where $v_p = \frac{trA}{n} - \gamma_p$ and

$$\gamma_p = \left[\frac{(n-1)^{2p-1}}{(n-1)^{2p-1}+1} tr\left(M(A) - \frac{trA}{n} I \right)^{2p} \right]^{1/2p}.$$

Proof In conjunction with Theorem 2.4 and Theorem 3.1, Theorem 3.2 immediately follows.

Remark 3.2 Set $p=1$ in (12) and note that

$$tr\left(A - \frac{trA}{n} I \right)^2 = tr\left(A^2\right) - \frac{(trA)^2}{n},$$

Theorem 3.2 yields into Theorem 2.5.

Therefore, if we do not know that an $n \times n$ M -matrix A has a real spectrum, we can construct an increasing sequence $\{v_p\}$ of lower bounds of $q(A)$ using the arithmetic symmetrization of the matrix A.

It is also important to keep in mind that we can deal with the power of 2 for p to increase the lower bounds of the minimum eigenvalue more quickly.

4. Numerical Examples

We consider two M-matrices.

Example 4.1 Let the symmetric M-matrix

$$A = \begin{pmatrix} 7 & -1 & -2 \\ -1 & 7 & -3 \\ -2 & -3 & 3 \end{pmatrix}.$$

The minimal eigenvalue of this matrix is $q(A) = 0.6122$. From (10), we acquire the lower bounds of $q(A)$ in Table 1.

Table 1. Lower bounds of $q(A)$

p	1	2	4	8
$q(A) \geq u_p$	0.5895	0.5895	0.6083	0.6122

However, Theorem 2.2 in [5] only provides the trivial lower bound of -0.5 for $q(A)$. It should be noted that the method of reference [4] does not apply to Example 4.1 since the matrix A does not satisfy the condition of weakly chained diagonally dominant.

Example 4.2 Let the nonsymmetric M-matrix

$$B = \begin{pmatrix} 4 & 0 & -2 \\ -1 & 4 & -1 \\ -1 & -1 & 2 \end{pmatrix}.$$

The minimal eigenvalue of this matrix is $q(B) = 0.8472$. From (12), we have the lower bounds for $q(B)$ in Table 2.

Table 2. Lower bounds of $q(B)$

p	1	2	4	8
$q(B) \geq v_p$	0.7947	0.7947	0.8041	0.8061

Given that the matrix is irreducibly diagonally dominant and thus weakly chained diagonally dominant, it is simple to demonstrate. From (2), we have $\|B^{-1}\|_{\infty} \leq 5/2$. So, by Theorem 2.1 (2), we get $q(B) \geq 2/5$. According to Theorem 2.2, we only obtain $q(B) \geq 0$. We see that in Example 4.1 and Example 4.2, our lower bounds give a better estimate than Theorem 2.1 and Theorem 2.2.

5. Conclusion

By building two increasing sequences, new lower bounds on the lowest eigenvalues of symmetric and nonsymmetric M-matrices are established. The methods can easily and tightly get better lower bounds. Two examples are provided to illustrate our methods by comparing them with the relevant conclusions.

Acknowledgments

We express our sincere thanks to the Science Foundation for Young teachers at Sichuan University Jinjiang College (QNJJ-2018-A01).

References

[1] R.A. Horn, C.R. Johnson, Topics in Matrix Analysis, *Cambridge University Press*, 1985.
[2] M. Özel, D. Varol, On the lower bounds for the minimum eigenvalue of the Hadamard product of an M-matrix and its inverse, *Georgian Math.*, Vol. 28, no. 3, pp. 439-443, 2021.
[3] A. Rastpour, J. Bourdeau-Marche, A Unified Treatment for the Inverses of M-Matrices and Scalars, *Mathematics Magazine*, pp. 1-4, 2022.
[4] P. N. Shivakumar, J. Joseph, Williams, Qiang Ye, et al, On two-side bounds related to weakly diagonally dominant M-matrices with applications to digital circuit dynamics, *SIAM J Matrix Anal Appl.*, Vol. 17, pp. 298-312,1996.
[5] W. Zhang, T. Z. Huang, Estimates of the minimal eigenvalue of irreducible M-matrices, *Chinese Journal of Engineering Mathematics*, Vol. 21, no. 8, pp. 31-34, 2004.
[6] O. Rojo, R. Soto, and H. Rojo, New eigenvalue estimates for complex matrices, *Computers Math. Applic.*, Vol. 25, no. 3, pp. 91-97, 1993.
[7] D. Taşçi, S. Kirkland, A sequence of upper bounds for the Perron root of a nonnegative matrix, *Linear Algebra Appl.*, Vol. 273, pp. 23-28, 1998.
[8] O. Rojo, R. Soto, and H. Rojo, A decreasing sequence of eigenvalue localization regions, *Linear Algebra Appl.*, Vol.196, pp. 71-84, 1994.

Convexity Estimates for the Solution of the Elliptical Partial Differential Equation $\det D^2u = e^u$

Xuemei YU [1]

[1]Beijing Polytechnic, 100176, China

Corresponding author and email: Tel.: 15727302001; email: 82321473@qq.com

Abstract: The Monge-Ampère equation $\det D^2u = e^u$ is completely nonlinear and elliptic, the convexity estimates for the solution of the elliptical partial differential equation is very important. We establish a differential inequality by constructing an auxiliary function and give two differential estimates for the solution of the equation $\det D^2u = e^u$ with 0 boundary value condition.

1. Introduction

The equation $\det D^2u = e^u$ is a special kind of Monge-Ampère equation. If the solution u of the equation is strictly convex, it is a completely nonlinear and it is an elliptical partial differential equation. The equation $\det D^2u = 1$ is studied in reference [1], they construct the auxiliary functions :

$$\varphi = \sum_{k,l=1}^{n} \frac{\partial \sigma_n(D^2u)}{\partial u_{kl}} u_k u_l - 2(n-1)u \text{ and } \psi = \sum_{k,l=1}^{n} \frac{\partial \sigma_2(D^2u)}{\partial u_{kl}} u_k u_l - 2u,$$

then get the estimations of mean curvature and the Gauss curvature.

For the class of completely nonlinear elliptic Monge-Ampère equations $\det D^2u = e^u$, D^2u is the Hessian matrix of u, the solution u is strictly convex when D^2u is positive definite [2]. We will study the elliptic Monge-Ampère equations $\det D^2u = e^u$ in 4-dimensional Euclidean spaces in this paper, which with 0 boundary value Dirichlet condition. The author gives the following results under the given conditions, the auxiliary functions

$$\varsigma = \sum_{k,l=1}^{4} \frac{\partial \sigma_4(D^2u)}{\partial u_{kl}} u_k u_l - 2\varepsilon u \text{ and } \eta = \sum_{k,l=1}^{4} \frac{\partial \sigma_2(D^2u)}{\partial u_{kl}} u_k u_l - 6\sqrt{\varepsilon} u$$

related to the strictly convex solution u is constructed to make it reach the maximum value at the boundary.

2. Preknowledge

2.1 Definition 2.1 [3]

Let $M=(m_{ij})$ be a symmetric $n$order matrix, $\lambda(M)=(\lambda_1(M),\lambda_2(M),\cdots,\lambda_n(M))$ is the eigenvalues of M. $\sigma_k(M)=\sigma_k(\lambda(M))$ is the sum of the k-order principal minor of the matrix M, especially $\sigma_n(m_{ij})=\det m_{ij}$.

2.2 Definition 2.2 [4]

If $\forall x\in\Omega$, $|\nabla u|\neq 0$. We call $\sum^{u(x_0)}=\{x\in\Omega\mid u(x)=u(x_0)\}$ is the level set of u at $x_0\in\Omega$.

2.3 Lemma 2.3 [5]

Let $\Omega\subset\mathbb{R}^n$ is a bounded connected domain and $\alpha_{ij}(x)$, $\beta_{ij}(x)$ are bounded continuous functions in $\overline{\Omega}$, for second order elliptic operators, $u\in C^2(\Omega)\cap C(\overline{\Omega})$, set

$$Lu=\sum_{i,j=1}^{n}\alpha_{ij}(x)u_{ij}+\sum_{i=1}^{n}\beta_{ij}(x)u_i,$$

u attains its maximum on $\partial\Omega$ when $Lu\geq 0$ in Ω.

2.4 Definition 2.4 [6]

We set:

$$a^{kl}=\frac{\partial\sigma_4(D^2u)}{\partial u_{kl}},\ b^{kl}=\frac{\partial\sigma_2(D^2u)}{\partial u_{kl}},$$

then

$$a^{kl}=u^{kl}\det(D^2u),\ b^{kl}=\begin{cases}\sum\limits_{\substack{j=1,\\ j\neq k}}^{4}u_{jj} & k=l\\ -u_{kl} & k\neq l\end{cases},$$

where $(u^{kl})=(u_{kl})^{-1}$.

3. Main theorems

3.1 Theorem 3.1

For $\forall x\in\Omega$, $\Omega\subset\mathbb{R}^4$, $u(x)$ is the solution of the equation

$$\begin{cases}\det D^2u=e^u & \text{in}\quad\Omega,\\ u=0 & \text{on}\quad\partial\Omega.\end{cases}\tag{1}$$

If u is strictly convex and Ω is a bounded convex, We have the following estimates:

$$\sum_{i,j=1}^{4}u^{ij}\varsigma_{ij}\geq 0,$$

where $\varsigma=\sum\limits_{k,l=1}^{4}\frac{\partial\sigma_4(D^2u)}{\partial u_{kl}}u_ku_l-2\varepsilon u$, $\varepsilon=e^{(\min\limits_{x\in\Omega}u)}$.

3.2 Theorem 3.2

For $\forall x \in \Omega$, $\Omega \subset \mathbb{R}^4$, $u(x)$ is the solution of the equation (1). If u is strictly convex and Ω is a bounded convex, we have the following estimates:

$$\sum_{i,j=1}^{4} u^{ij}\eta_{ij} \geq 0,$$

where $\eta = \sum_{k,l=1}^{4} \frac{\partial \sigma_2(D^2u)}{\partial u_{kl}} u_k u_l - 6\sqrt{\varepsilon u}$, $\varepsilon = e^{(\min_{x\in\Omega} u)}$.

4. Proof of theorems

In order to prove the estimates is true at any point $x_0 \in \Omega$. We choose the coordinates e_1, e_2, e_3, e_4 for any $x_0 \in \Omega$ such that the Hessian matrix $u_{ij}(x_0)$ of u is diagonal [7]. The following proof will be done at x_0.

Calculate the first derivative ς_i.

$$\begin{aligned}
\varsigma_i &= \Big(\sum_{k,l=1}^{4} e^u u^{kl} u_k u_l\Big)_i - 2\varepsilon u_i \\
&= \sum_{k,l=1}^{4} (e^u u^{kl}{}_i u_k u_l + 2e^u u^{kl} u_{ki} u_l + e^u u^{kl} u_k u_l u_i) - 2\varepsilon u_i \\
&= \sum_{k,l=1}^{4} (e^u u^{kl}{}_i u_k u_l + e^u u^{kl} u_k u_l u_i) + 2e^u \sum_{k,l=1}^{4} u^{kl} u_{ki} u_l - 2\varepsilon u_i\,, \\
&= \sum_{k,l=1}^{4} (e^u u^{kl}{}_i u_k u_l + e^u u^{kl} u_k u_l u_i) + 2e^u u_i - 2\varepsilon u_i \\
&= \sum_{k,l=1}^{4} (e^u u^{kl}{}_i u_k u_l + e^u u^{kl} u_k u_l u_i) + 2(e^u - \varepsilon) u_i
\end{aligned}$$

Let's go on to the second derivative,

$$\begin{aligned}
\varsigma_{ii} &= \sum_{k,l=1}^{4} (e^u u^{kl}{}_{ii} u_k u_l + 2e^u u^{kl}{}_i u_{ki} u_l + 2e^u u^{kl}{}_i u_k u_l u_i \\
&\quad + e^u u^{kl} u_k u_l u_i^2 + 2e^u u^{kl} u_{ki} u_l u_i + e^u u^{kl} u_k u_l u_{ii}) \\
&\quad + 2e^u u_i^2 + 2(e^u - \varepsilon) u_{ii} \\
&= \sum_{k,l=1}^{4} e^u u^{kl}{}_{ii} u_k u_l + 2\sum_{l=1}^{4} e^u u^{il}{}_i u_{ii} u_l + 2\sum_{l=1}^{4} e^u u^{il}{}_i u_l u_i^2 \\
&\quad + \sum_{k=1}^{4} e^u u^{kk} u_k^2 u_i^2 + 5e^u u_i^2 + 2(e^u - \varepsilon) u_{ii}
\end{aligned}$$

Thus,

$$\sum_{i,j=1}^{4} u^{ij}\varsigma_{ij} = \sum_{i=1}^{4} u^{ii}\varsigma_{ii}$$
$$= \sum_{k,l,i=1}^{4} e^{u}u^{kl}{}_{ii}u_k u_l u^{ii} + 2\sum_{i,l=1}^{4}(e^{u}u^{il}{}_{i}u_{ii}u_l u^{ii} + e^{u}u^{il}{}_{i}u_l u_i^2 u^{ii}) \tag{2}$$
$$+ \sum_{i,k=1}^{4} e^{u}u^{kk}u_k^2 u_i^2 u^{ii} + 5\sum_{i=1}^{3} e^{u}u_i^2 u^{ii} + 8(e^{u} - \varepsilon)$$

Because of $\det D^2 u = e^u$, differential calculations are performed on both sides at the same, we can get

$$\frac{\partial \det D^2 u}{\partial x_k} = \det D^2 u \sum_{i,j=1}^{4} u^{ij}u_{ijk} = e^{u}u_k ,$$

i.e.

$$\sum_{i=1}^{4} u^{ii}u_{iik} = u_k . \tag{3}$$

Differential $\det D^2 u = e^u$ again,

$$\sum_{i,j=1}^{4} u^{ij}u_{ijkl} - \sum_{i,j,p,q=1}^{4} u^{iq}u^{pj}u_{ijk}u_{pql} = u_{kl}$$

$$\sum_{i=1}^{4} u^{ii}u_{iikl} = \sum_{i,j=1}^{4} u^{ii}u^{jj}u_{ijk}u_{ijl} + u_{kl} . \tag{4}$$

Apply

$$u^{kl}{}_{i} = -\sum_{p,q=1}^{4} u^{kq}u^{pl}u_{pqi} \tag{5}$$

to get

$$u^{kl}{}_{ii} = (-\sum_{p,q=1}^{4} u^{kq}u^{pl}u_{pqi})_i$$
$$= \sum_{p,q,m,n=1}^{4} (u^{kn}u^{mq}u^{pl} + u^{kq}u^{pn}u^{ml})u_{pqi}u_{mni} - \sum_{p,q=1}^{4} u^{kq}u^{pl}u_{pqii} \tag{6}$$
$$= 2\sum_{j=1}^{4} u^{kk}u^{ll}u^{jj}u_{jki}u_{jli} - u^{kk}u^{ll}u_{klii}$$

Substituting (3)(4)(5)(6) into (2), we can get

$$\sum_{i,j=1}^{4} u^{ij}\varsigma_{ij} = \sum_{\substack{i,j,k,l=1,\\ k\neq l}}^{4} e^{u}u^{ii}u^{jj}(u^{kk}u_k u_{ijk})^2 - \sum_{\substack{k,l=1,\\ k\neq l}}^{4} e^{u}u^{kk}u^{ll}u_{kl}u_k u_l + \sum_{i,j=1}^{4} e^{u}u^{ii}u_i^2 u^{jj}u_j^2$$
$$-2\sum_{i,j,k=1}^{4} e^{u}u^{ii}u^{jj}u^{kk}u_i u_j u_k u_{ijk} + 2\sum_{i=1}^{4} e^{u}u_i^2 u^{ii} + 8(e^{u} - \varepsilon)$$
$$= e^{u}\sum_{i,j=1}^{4} u^{ii}u^{jj}(\sum_{k=1}^{4} u^{kk}u_k u_{ijk} - u_i u_j)^2 + 2e^{u}\sum_{i=1}^{4} u_i^2 u^{ii} + 8(e^{u} - \varepsilon)$$

Because $e^u > 0$, $\varepsilon = e^{(\min_{x\in\Omega} u)}$, we get the differential estimates $\sum_{i,j=1}^{4} u^{ij}\varsigma_{ij} \geq 0$.

Under the same assumptions, we can get the function

$$\eta = \sum_{k,l=1}^{4} \frac{\partial \sigma_2(D^2u)}{\partial u_{kl}} u_k u_l - 6\sqrt{\varepsilon} u = \sum_{k,l=1}^{4} b^{kl} u_k u_l - 6\sqrt{\varepsilon} u$$

from Definition 2.4. Calculate the first derivative of η ; we get

$$\begin{aligned}\eta_i &= (\sum_{k,l=1}^{4} b^{kl} u_k u_l)_i - 6\sqrt{\varepsilon} u_i \\ &= \sum_{k,l=1}^{4} b^{kl}{}_i u_k u_l + \sum_{k,l=1}^{4} b^{kl} u_{ki} u_l + \sum_{k,l=1}^{4} b^{kl} u_k u_{li} - 6\sqrt{\varepsilon} u_i , \\ &= \sum_{k,l=1}^{4} b^{kl}{}_i u_k u_l + 2\sum_{k,l=1}^{4} b^{kl} u_{ki} u_l - 6\sqrt{\varepsilon} u_i\end{aligned}$$

Calculate the second derivative again,

$$\begin{aligned}\eta_{ii} &= \sum_{k,l=1}^{4} (b^{kl}{}_{ii} u_k u_l + 2b^{kl}{}_i u_{ki} u_l) + \sum_{k,l=1}^{4} (2b^{kl}{}_i u_{ki} u_l + 2b^{kl} u_{kii} u_l + 2b^{kl} u_{ki} u_{li}) - 6\sqrt{\varepsilon} u_{ii} \\ &= \sum_{k,l=1}^{4} b^{kl}{}_{ii} u_k u_l + 4\sum_{k,l=1}^{4} b^{kl}{}_i u_{ki} u_l + 2\sum_{k,l=1}^{4} b^{kl} u_{kii} u_l + 2\sum_{k,l=1}^{4} b^{kl} u_{ki} u_{li} - 6\sqrt{\varepsilon} u_{ii} \\ &= \sum_{k,l=1}^{4} b^{kl}{}_{ii} u_k u_l + 4\sum_{l=1}^{4} b^{il}{}_i u_{ii} u_l + 2\sum_{k=1}^{4} b^{kk} u_{kii} u_k + 2b^{ii} u_{ii}^2 - 6\sqrt{\varepsilon} u_{ii}\end{aligned}$$

Thus,

$$\sum_{i,j=1}^{4} u^{ij}\eta_{ij} = \sum_{i=1}^{4} u^{ii}\eta_{ii} = \sum_{k,l,i=1}^{4} b^{kl}{}_{ii} u_k u_l u^{ii} + 4\sum_{i,l=1}^{4} b^{il}{}_i u_{ii} u_l u^{ii} + 2\sum_{i,k=1}^{4} b^{kk} u_{kii} u_k u^{ii} + 2\sum_{i=1}^{4} b^{ii} u_{ii} - 24\sqrt{\varepsilon}$$

Because of $\sum_{i=1}^{4} u^{ii} u_{iik} = u_k$, $\sum_{i=1}^{4} b^{il}{}_i = 0$, $\sum_{i=1}^{4} b^{ii} u_{ii} = 2\sigma_2(D^2u)$, we can get

$$\sum_{i,j=1}^{4} u^{ij}\eta_{ij} = \sum_{k,l,i=1}^{4} b^{kl}{}_{ii} u_k u_l u^{ii} + 2\sum_{i,k=1}^{4} b^{kk} u_k^2 + 4\sigma_2(D^2u) - 24\sqrt{\varepsilon} \tag{7}$$

And because

$$b^{kl} = \begin{cases} \sum_{j=1, j\neq k}^{4} u_{jj} & k = l \\ -u_{kl} & k \neq l \end{cases}, \quad b^{kl}{}_{ii} = \begin{cases} \sum_{j=1, j\neq k}^{4} u_{jjii} & k = l \\ -u_{klii} & k \neq l \end{cases}$$

We know

$$\sum_{k,l,i=1}^{4} b^{kl}{}_{ii} u_k u_l u^{ii} = \sum_{i,j,l=1, j\neq l}^{4} u^{ii} u_{iijj} u_l^2 - \sum_{i,k,l=1, k\neq l}^{4} u^{ii} u_{iikl} u_k u_l ,$$

Applying (4), we obtain

$$\begin{aligned}\sum_{k,l,i=1}^{4} b^{kl}{}_{ii}u_k u_l u^{ii} &= \sum_{\substack{i,j,k,l=1,\\ k\neq l}}^{4} u^{ii}\, u^{jj}u_{ijk}^2 u_l^2 + \sum_{\substack{k,l=1,\\ k\neq l}}^{4} u_{kl}u_k u_l \\ &\quad - \sum_{\substack{i,j,k,l=1,\\ k\neq l}}^{4} u^{ii}\, u^{jj}u_{ijk}u_{ijl}u_k u_l - \sum_{\substack{i,k,l=1,\\ k\neq l}}^{4} u_{kk}u_l^2 \\ &= \sum_{\substack{i,j,k,l=1,\\ k\neq l}}^{4} (u^{ii}\, u^{jj}u_{ijk}^2 u_l^2 - u^{ii}\, u^{jj}u_{ijk}u_{ijl}u_k u_l) - \sum_{\substack{i,k,l=1,\\ k\neq l}}^{4} u_{kk}u_l^2\end{aligned} \tag{8}$$

Substituting (8) into (7), we obtain

$$\sum_{i,j=1}^{4} u^{ij}\eta_{ij} = \sum_{i,j=1}^{4}[u^{ii}\, u^{jj}\sum_{k,l=1}^{4}(u_{ijk}^2 u_l^2 - u_{ijk}u_{ijl}u_k u_l)] + \sum_{i,k,l=1,\ k\neq l}^{4} u_{kk}u_l^2 + 4\sigma_2(D^2u) - 24\sqrt{\varepsilon}$$

where

$$\sum_{k,l=1,\ k\neq l}^{4} u_{kk}u_l^2 = \sum_{k=1}^{4} b^{kk}u_k^2 > 0,$$

Set

$$S_k = \frac{\sigma_k}{C_n^k},\ \sigma_0 = 1,$$

Then

$$S_{k-1}S_{k+1} \le S_k^2, k = 1,2,\ldots,n-1.$$

$$S_1 \ge S_2^{\frac{1}{2}} \ge \cdots S_n^{\frac{1}{n}}, k = 1,2,\ldots,n-1.$$

We know

$$\sigma_2(D^2u) \ge C_4^2(\sigma_4(D^2u))^{\frac{1}{2}} = 6\sqrt{e^u} \ge 6\sqrt{\varepsilon} \tag{9}$$

by Newton's inequality.

We know

$$\sum_{k,l=1}^{4}(u_{ijk}^2 u_l^2 - u_{ijk}u_{ijl}u_k u_l) \ge 0 \tag{10}$$

by Cauchy-Schwarz's inequality.

Combining (9) and (10), the differential estimates $\sum_{i,j=1}^{4} u^{ij}\eta_{ij} \ge 0$ is got.

5. Conclusions

We get two estimates of the solution u of the equation $\det D^2u = e^u$ in this paper. We chose the auxiliary functions in bounded convex domains in four-dimensional Euclidean spaces, then get two differential inequalities. These auxiliary functions are related to curvature of the level sets of the solutions. These differential estimates research contents are helpful for further study of convex solutions, for example the mean curvature and the Gauss curvature.

Acknowledgment

Fund project "The convexity estimate of the level set of solutions for a class of Monge-Ampère equations" (NO. 2022X015-KXZ).

Journal of Physics: Conference Series **2455** (2023) 012016 doi:10.1088/1742-6596/2455/1/012016

References

[1] Chen C Q, Ma X N, Shi S J, Curvature Estimates for the Level Sets of Solutions of the Monge-Ampère Equation $\det D^2 u = 1$, Chinese Annals of Mathematics, 2014 (35B): 895-906.

[2] Chen C Q, Shi S J, Curvature Estimates for the Level Sets of Spatial Quasiconcave Solutions to A Class of Parabolic Equations, Sci. China Math.,2011,54 (10),2063-2080.

[3] Yu X M. The mean curvature estimate for the level sets of solutions of the Monge-Ampère equation on Riemannian manifold, Journal of Physics:Conference Series, 2021.

[4] Ma X N, Zhang Y B. The Convexity and the Gaussian Curvature Estimates for the Level Sets of Harmonic Functions on Convex Rings in Space Forms. J.Geom. Anal..2014,24:340-341.

[5] Feng H Y, Shi S J. Curvature estimates for the level sets of solutions to a class of Monge–Ampère equations, NonlinearAnalysis, 2019.

[6] Yu X M. A Differential Inequality for the Solution of Monge Ampère Equation on Riemannian Manifold", Journal of Physics: Conference Series, 2021.

[7] Feng H Y. The convexity estimates of solutions for a class of elliptic Hessian equations. Master's thesis in Harbin Normal University. 2017.

A Numerical Study of the Poisson Equation with Arbitrary Domains Based on the Extrapolation Technique

Hui Zhang*, Pan Yu*, Xueying Shi*, Lin Li*,†

*School of Mathematics and Physics and Hunan Key Laboratory for Mathematical Modeling and Scientific Computing, University of South China, Hengyang, 421001, China.

†Corresponding author: lilinmath@usc.edu.cn.

Abstract. The Poisson equation with arbitrary domains is considered in this paper using an extrapolation technique. The two-dimensional problem is primarily examined, and the extrapolation formulas O2 and O4 are thoroughly deduced. Not only the effect of different extrapolations on boundary point errors is analyzed, but a strategy for selecting extrapolation points is also provided. Finally, numerical accuracy results are presented.

1. Introduction

The numerical solution of Poisson's equation is a difficult problem, and people often transfer the solution in a regular region to the solution in an irregular region. For solving partial differential equations, the spectral method, finite volume method, and finite-difference method are now the most popular methods.

Spectral approaches often do not require a large number of terms to produce high-precision solutions to problems in two or more dimensions. However, J.P.Boyd et al. pointed out in [2] that when dealing with complicated geometric models and discontinuous interfaces, the spectral method's solution accuracy will be severely harmed.

Although the finite volume technique can solve the flaws of the Taylor expansion, Li pointed out in [3] that studying the mathematical properties of the equation with this method is inconvenient.

The finite difference method is a simple and effective method, [5] which makes reference to it. Higher-order schemes are very easy to obtain on regular grids. On the other hand, if complicated flows are defined, the constraint on simple geometry is a significant disadvantage.

Finite difference and multigrid techniques are commonly employed to solve partial differential equations in complicated domains. Multigrid provides a lot of advantages when it comes to expressing numerical calculations in a complicated domain. The multigrid algorithm has been widely utilized in the computation of difficult geometry issues since literature [1]. These techniques are based on the Cartesian coordinate system, which is coupled with various boundary processing methods to provide distinct results.

The Monte Carlo Markov chain approach has been employed by Zheng et al. [19] to solve the Poisson equation with the irregular area. Meanwhile, He et al. [8] used the Trefftz approach to consider the Poisson equation. Although not simple to compute, this approach is successful. The solution to the Poisson equation for two-dimensional irregular areas will thus be the main emphasis of this work [17].To establish an outgoing general formula to solve the irregular area using finite differences and extrapolation techniques, this study applies Lagrange interpolation. The Poisson equation's general formula is:

$$\Delta u = f. \tag{1}$$

We use the virtual boundary approach given by Saiki and Biringen in their paper [14] since extrapolated values must be established on virtual components. The solution of the Poisson equation in two dimensions is the sole focus of this work. In two-dimensional, we add a deletion factor, and the deletion factor's role is to bring the extrapolation point closer to the boundary, lowering the effect of extrapolation error. This improves the accuracy of the equation's answer.

The rest of the paper is organized as follows: section 2 describes the Lagrange interpolation method and the reasons for introducing the extrapolation method, as well as the extrapolation formula. Section 3 introduces an extrapolation point decision scheme for two-dimensional problems, which we apply to the clover model to provide comprehensive error analysis and accuracy. Section 4 contains the conclusions.

2. Extrapolation Based on Lagrange Interpolation

We cannot guarantee that every boundary point within an irregular region is also the grid's division point. Therefore, to solve Poisson equations with constant coefficients in the irregular region, we employ the finite difference approach and the extrapolation method. The extrapolation method is derived from the Lagrange interpolation method. In [7], Gibou et al. discussed three different types of extrapolations. The extrapolation formula's derivation will be briefly given here.

2.1. Lagrange interpolation method.

The Lagrange interpolation method [15] is a method of fitting the interpolation polynomial by using the inherent law of the problem. This paper uses the extrapolation method derived from the Lagrange interpolation method $L_n(x)$, solving the numerical problems of nonuniform grids in two dimensions,

$$L_n(x) = \sum_{i=0}^{n} u_i l_i(x), \tag{2}$$

Where $l_i(x)$ is as follows:

$$l_i(x) = \prod_{j=0}^{n} \frac{x-x_j}{x_i-x_j}. \tag{3}$$

However, since the Lagrange interpolation method is an approximate fitting function method, there must be an error $R_n(x)$,

$$R_n(x) \equiv f(x) - L_n(x) = \frac{f^{n+1}(\xi)}{(n+1)!}\omega_{n+1}(x). \tag{4}$$

Where ξ belongs to the grid area and depends on x,

$$\omega_{n+1}(x) = \prod_{j=0}^{n} x - x_j. \tag{5}$$

Remark 1: We can also use the finite difference method, which is described in full in [12], to calculate the mesh generation issue. It is easy to see that the error phenomena will diminish as the number of grid points rises. However, with the Lagrange interpolation approach, the associated error effect may be more inaccurate the higher the corresponding interpolation number. For instance, when some functions' high-order Lagrange interpolation polynomial interpolation is built using uniform nodes, the Runge phenomenon [4] refers to the phenomenon that the error at the edge of the interpolation interval may be considerable. Therefore, we shall introduce the extrapolation approach for the analysis of numerical issues.

2.2. Extrapolation method.

When we mesh any region, there are two cases: one is to mesh the regular region, which can find at least one finite difference method [13] to make all points on the boundary fall on the center of the grid. The other is the grid division of irregular regions, which cannot find any finite difference method to make all points on the boundary fall at the center of the grid. If a finite difference method is used for the latter, the error increases because the boundary points that do not fall on the grid center do not have actual values. To reduce the error increase, we will introduce another method: the extrapolation method based on the central finite difference method. For example, O2 and O4 [18] are respectively expressed as second-order and fourth-order central finite difference methods, which are used to calculate the grid points of internal rules; B0, B1, B2, and B3 are respectively expressed as constant, linear, quadratic, and cubic extrapolation methods, and they are used to calculate the extrapolation.

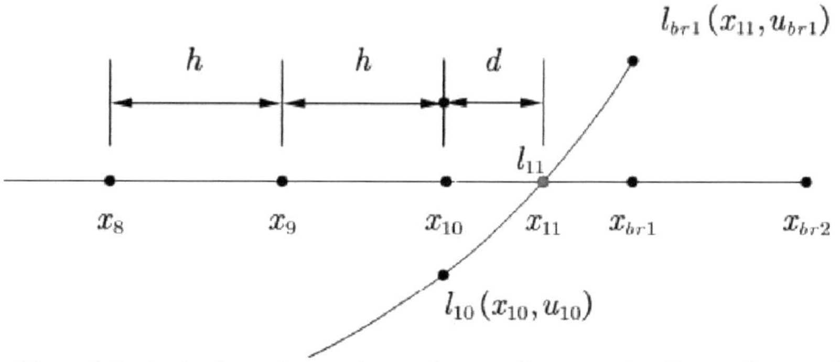

FIGURE 1. The red dot is the boundary point, and x_{br1} and x_{br2} are the first and second extrapolation points, respectively.

In a particular area, we set the grid number $N = 10$. The calculations for E_1 and E_{10} will require extrapolations if the calculation element is expressed as E_i, $i = 1,2, \ldots,10$. Let's write the coordinate value pair of position and variable as $l_i\ (x_i, u_i)$,$i = 1,2, \ldots,11$, including extrapolation points. The left extrapolation point is $l_{bl}\,(x_{bl}, u_{bl})$, and the right extrapolation point is $l_{br}\,(x_{br}, u_{br})$. Let us only describe the right boundary, similar definitions apply to the left boundary. Next, we will focus on the specific derivation process of extrapolation.

Based on O2, the specific formulas of the extrapolation method are derived by the Lagrange interpolation method as follows:

Constant extrapolation. Extrapolation equals boundary value: $u_{br} = u_{11}$.

Linear extrapolation. Where $h = x_{br} - x_{10}, d = x_{11} - x_{10}, \theta = \frac{d}{h}$, the following formula is obtained by the Lagrange interpolation method

$$L_1(x) = \frac{x-x_{10}}{x_{11}-x_{10}} u_{11} + \frac{x-x_{11}}{x_{10}-x_{11}} u_{10}. \tag{6}$$

Let us calculate u_{br} by making $x = x_{br}$,

$$L_1(x_{br}) = u_{br} = \frac{x_{br}-x_{10}}{x_{11}-x_{10}} u_{11} + \frac{x_{br}-x_{11}}{x_{10}-x_{11}} u_{10}. \tag{7}$$

Then, the extrapolation value u_{br} is calculated by substitution:

$$u_{br} = \frac{1}{\theta} u_{11} + \frac{\theta-1}{\theta} u_{10}. \tag{8}$$

Quadratic extrapolation. The value u_{br} of quadratic extrapolation is calculated by the same method as a linear extrapolation

$$u_{br} = \frac{\theta(1-\theta)u_9 - 2(1-\theta^2)u_{10} + u_{11}}{\theta(1+\theta)}. \tag{9}$$

Cubic extrapolation. Omitting the derivation steps, we only give the results of the extrapolated values u_{br}.

$$u_{br} = \frac{6}{\theta(\theta+1)(\theta+2)} u_{11} + \frac{3(\theta-1)}{\theta} u_{10} + \frac{3(1-\theta)}{\theta+1} u_9 + \frac{\theta-1}{\theta+2} u_8. \tag{10}$$

In the case of O4, the values of the first extrapolation point of B1 and B2 have been solved by the Lagrange interpolation method. If $h = x_{br2} - x_{br1}$, we can calculate u_{br2} using a similar Lagrange interpolation method. A similar approach can be used on the left. Constant extrapolation: $u_{br2} = u_{11}$, $u_{bl2} = u_1$. The expressions of other extrapolation methods are shown in Table 1 below:

TABLE 1. The summary of the extrapolation formulas for O4.

	u_{br2}	u_{bl2}
B1	$\frac{(\theta-2)u_{br1}+u_{11}}{\theta-1}$	$\frac{(\theta-2)u_{bl1}+u_1}{\theta-1}$
B2	$\frac{2(\theta-2)}{\theta-1}u_{br1}+\frac{2}{\theta(\theta-1)}u_{11}+\frac{2-\theta}{\theta}u_{10}$	$\frac{2(\theta-2)}{\theta-1}u_{bl1}+\frac{2}{\theta(\theta-1)}u_1+\frac{2-\theta}{\theta}u_2$
B3	$\frac{6-3\theta}{1-\theta}u_{br1}+\frac{6}{\theta^3-\theta}u_{11}+\frac{6-3\theta}{\theta}u_{10}+\frac{\theta-2}{1-\theta}u_9$	$\frac{6-3\theta}{1-\theta}u_{bl1}+\frac{6}{\theta^3-\theta}u_1+\frac{6-3\theta}{\theta}u_2+\frac{\theta-2}{1-\theta}u_3$

The two-dimensional problem is analyzed and discussed in the sections that follow, with a focus on O2-B0, O2-B1, O2-B2, and O4-B0, and it is shown whether our solution is workable in light of the conclusion.

3. Numerical experiments

The two-dimensional constant coefficient equation is the following Poisson equation:

$$u_{xx} + u_{yy} = f. \qquad (11)$$

Where Δx, Δy are the length of a regular grid, we set $\Delta x = \Delta y = h$. According to the related articles of Fukuchi [6] [16] [17], we know that there are many methods to solve the two-dimensional Poisson equation, and using only finite difference and extrapolation is one of them. We compare other methods with this method to analyze its solution effect. Here we only discuss the solution of the following two cases.

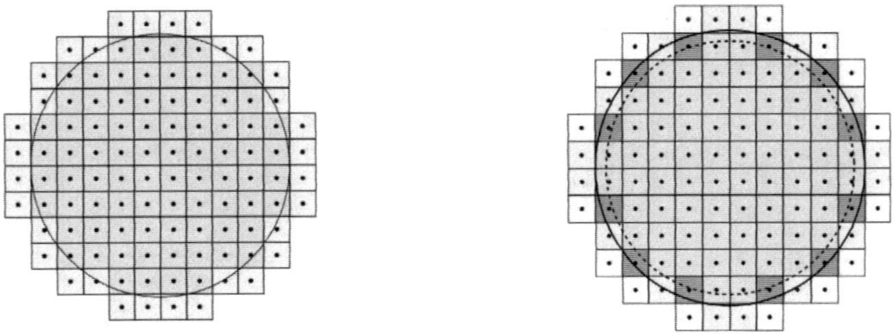

(a) Case 1 (b) Case 2

FIGURE 2. Cyan points are extrapolation points and gray points are interior points. Case 1: divide extrapolation points and interior points by graphic boundary; Case 2: based on case one, the delete factor is added.

Figure 2(a) shows Case 1. First, we specify that the values of the center points of all grids be taken to represent the area values of the grid. All Grid points are divided into two parts according to the location of the Grid Center Point. If a grid point is within an irregular area, it is called an interior point. If the grid point is outside the irregular area, it is called the outer point. In an outer point, all points adjacent to an inner point are called extrapolation points.

Figure 2(b) shows Case 2. The Yellow grid refers to the grid that is the extrapolation point in Case 1, but now is not the extrapolation point and does not participate in the subsequent calculation. The purple grid means that the inner point of Case 1 is now an extrapolation point. In this case, the deletion factor ω is added. For the above deletion factor ω, we will explain and analyze it in detail. The deletion factor is the distance between the center of the internal element close to the boundary and the boundary, and its value range is $0 \leq \omega \leq 1$. As shown in Figure 2(a)-Case 1, we assume that the grid configuration unit is $\triangle y$, and we can find that the center and boundary of internal elements close to the boundary are 0.5 $\triangle y$. Then we delete the grid to obtain a two-dimensional grid diagram after the action of the deletion factor ω in Figure 2(b)-Case 2, and its extrapolation elements, boundary elements, and internal elements have also changed accordingly. The reason why the deletion factor is introduced is to narrow the range of complex graphics and reduce the error caused by complex calculations in the future. Naturally, the value of the deletion factor cannot be selected at will. There is a reasonable range for each model. When the value is not within the range, the calculation error will increase.

In the two-dimensional grid calculation, when using the extrapolation method with B1 and above accuracy, it is necessary to solve the problem that one boundary point corresponds to several extrapolation formulas. For example, as shown in Figure 3, the finite difference method using O2 at point u_5 has:

$$\frac{u_4-2u_5+u_6}{\triangle x^2}+\frac{u_2-2u_5+u_8}{\triangle y^2}=f. \tag{12}$$

At this time, the extrapolation point u_6 is involved, and u_6 is the extrapolation point of u_9, so u_6 corresponds to the values of the two extrapolation formulas. If two formulas are used at the same time, u_6 will have two extrapolated values at the same time, resulting in no solution to the whole equation system. Therefore, we only select one of the extrapolated values to represent u_6. For example, here we use the B1 method (8) of u_5 to deduce the value of u_6:

$$u_6=\frac{1}{\theta}u^*+\frac{\theta-1}{\theta}u_5, \tag{13}$$

Where u^* is the boundary value. Then, when calculating the value of u_9, (13) is directly substituted for calculation. Using this calculation method will also produce corresponding errors. When the selected extrapolation values are different, the results may be slightly different. To avoid the influence of this error on our research, we specify the selection method: When the extrapolation point is the extrapolation of two interior points, select the extrapolation formula corresponding to the left or right extrapolation point. We uniformly set f = -1 to calculate and analyze the graphics in irregular areas.

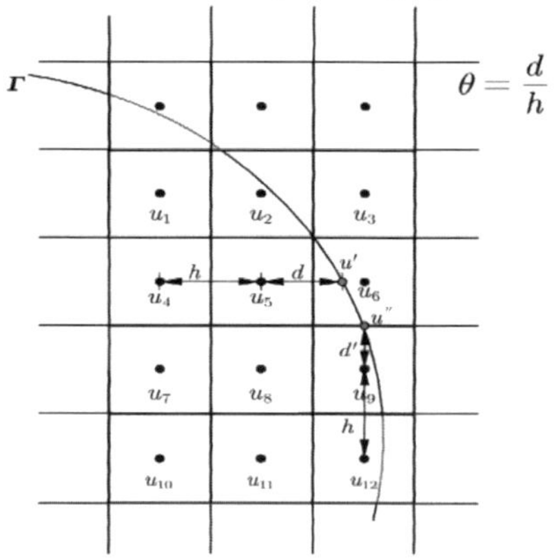

FIGURE 3. Boundary partial graph.

Next, we made a theoretical study on the shape of the clover. According to the two cases of Figure 2, the cloverleaf model is divided into grids. The effect of solving the equation is reflected by the calculation error. The solution errors of the two methods are compared and analyzed. The equation for clover is

$$r = 2 + \sin(3\theta), \qquad \theta \in [0,2\pi) \tag{14}$$

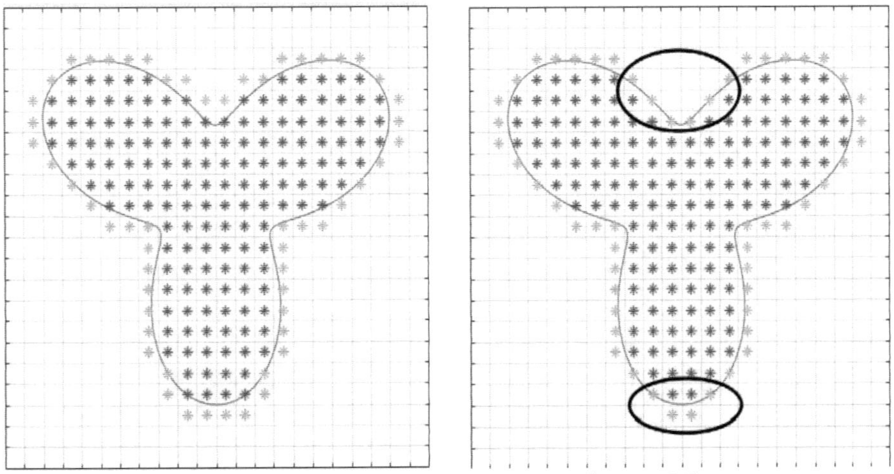

(a)$\omega = 0$ (b)$\omega = 0.5$

FIGURE 4. The above method is used to segment the graph. (a) $\triangle x = \triangle y = 0.1, \omega = 0$. (b) $\triangle x = \triangle y = 0.1, \omega = 0.5$.

The schematic diagram of graph meshing is presented in Figure 4, where Γ is the boundary of the graphical, and $u(\Gamma) = 0$. The red point in the drawing is the internal point, and the green point is the extrapolation point. Figure 4(b) uses the result of Figure 2. It can be seen that the classification of points in the circled places is different. When using the extrapolation method of more than one order, a problem needs to be solved. It can be seen from the figure that some extrapolation points are adjacent to two interior points, which means that this extrapolation point has two extrapolation formulas, which will make the equations have no solution. Therefore, in this case, we only choose one extrapolation formula to substitute into the equations for the solution. There will be the same problem when using the FD method of O4. The detailed errors after solving the two cases are shown in the table below.

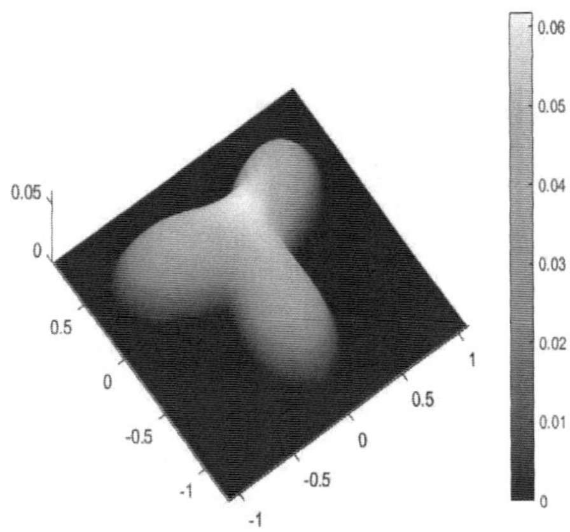

FIGURE 5. The solution in this case: $\triangle x = \triangle y = 0.025, \omega = 0.5$.

TABLE 2. Calculation error, $\Omega_x = [-1.1,1.1]$, $\Omega_y = [-1.3,0.9]$, $u(\Gamma) = 0$, $h = 0.1$.

h	O2-B0	O2-B1	O2-B2	O4-B0
h	8.26E-02	6.67E-02	6.69E-02	6.90E-02
0.5h	3.93E-02	3.60E-02	3.65E-02	3.12E-02
0.25h	1.37E-02	7.96E-03	8.24E-03	1.02E-02
h/0.5h	2.10	2.14	2.14	2.21
0.5h/0.25h	2.86	4.52	4.43	3.05
AP	1	2	2	1

TABLE 3. Calculation error, $\Omega_x = [-1.1,1.1]$, $\Omega_y = [-1.3,0.9]$, $u(\Gamma) = 0$, $h = 0.1$, $\omega = 0.5$.

h	O2-B0	O2-B1	O2-B2	O4-B0
h	7.10E-02	5.76E-02	5.70E-02	5.94E-02
0.5h	3.75E-02	2.88E-02	3.16E-02	1.35E-02
0.25h	1.26E-02	1.05E-03	7.75E-03	3.32E-03
h/0.5h	1.89	2.00	1.80	4.40
0.5h/0.25h	2.97	4.04	4.08	4.07
AP	1	2	2	2

By comparing the error data in Table 2 and Table 3, it can be obviously found that the error after adding the deletion factor is significantly reduced. When the extrapolation error increases, B0 to B1 decreases significantly but increases after B1 to B2. Because the order of the exact solution itself is one dimension, the use of high-order extrapolation will increase the error. When the order of finite difference is increased, not only the error is reduced but also the accuracy is improved quickly.

4. Conclusion

In this article, we primarily focus on the Poisson equation's numerical solution with a two-dimensional irregular region. We offer our algorithm, which was inspired by [9] [10] [11]. The process involves meshing the problem area first, using the extrapolation method for the boundary point and the finite difference approach for the center point, and then showcasing the algorithm's benefits and drawbacks through changes in error. We summarize the impact of the algorithm on the equation's solution after solving the aforementioned model. The following are the main points:

while adding more meshes can reduce error, it cannot always increase the correctness of the result. The extrapolation method's influence on the equation's solution will also be lessened as the number of grids increases, leading to nearly identical computation results for the B1 and B2 approaches.

The accuracy of the approach utilized, which is mostly influenced by the accuracy of the finite difference method, is related to the accuracy of the solution. Even if the extrapolation method's accuracy improves, when O2 or O4 are utilized, the accuracy of the solution won't go above 3 or 5.

In summary, our algorithm can effectively solve the Poisson equation with arbitrary regions, and it may even be extended to other problems.

Acknowledgment.

This work is partially supported by the Science Foundations of Hunan Province (Nos:2020JJ5464, 20C1595)

References

[1] M. J. Aftosmis. Robust and efficient cartesian mesh generation for component-based geometry. *AIAA Journal*, 36(6):952–960, 1997.

[2] J. P. Boyd and T. Marilyn. Chebyshev and Fourier spectral methods. *Dover Publications*, 2(1-2):19–24, 2000.

[3] L. Chen, K. K. Zhou, and Z. X. Li. Dynamic characteristic fitting of air spring and calculation and analysis of variable stiffness of air suspension. *Journal of Mechanical Engineering*, 1(4):6, 2010.

[4] Y. J. Chen, H. Y. He, and S. L. Zhang. A new algebra interpolation polynomial without runge phenomenon. *Applied Mechanics and Materials*, 23(8):303–306, 2013.

[5] G. E. Forsythe and W. R. Wasow. Finite difference methods for partial differential equations. *Shanghai Science and Technology Press*, 1964.

[6] T. Fukuchi. Numerical analysis of hagen-poiseuille flow using square calculation elements. In *Nctam Papers, National Congress of Theoretical Applied Mechanics, Japan*, 2004.

[7] F. Gibou, R. P. Fedkiw, L. T. Cheng, and M. Kang. A second-order-accurate symmetric discretization of the poisson equation on irregular domains. *Journal of Computational Physics*, 176(1):205–227, 2002.

[8] X. S. He and C. J. Wu. Solving poisson equation based on indirect trefftz method. *Proceedings of the 2009 Ship Structural Mechanics Academic Conference and the 30th Anniversary Commemorative Conference of Chinese Ship Academia Entering ISSC*, 2009.

[9] Z. Jomaa and C. Macaskill. The embedded finite difference method for the poisson equation in a domain with an irregular boundary and dirichlet boundary conditions. *Journal of Computational Physics*, 202(2):488–506, 2005.

[10] Z. Jomaa and C. Macaskill. Numerical solution of the 2d poisson equation on an irregular domain with robin boundary conditions. *Australian New Zealand Industrial Applied Mathematics Journal*, 50(4):1–20, 2008.

[11] Z. Jomaa and C. Macaskill. The shortley–weller embedded finite-difference method for the 3d poisson equation with mixed boundary conditions. *Journal of Computational Physics*, 229(10):3675–3690, 2010.

[12] J. Peir´o and S. Sherwin. *Finite Difference, Finite Element and Finite Volume Methods for Partial Differential Equations*. Handbook of Materials Modeling, 2005.

[13] L. I. Qingyang, W. U. Guochen, and P. Duan. Elastic wavefield forward modeling in heterogeneous media based on the quasi-regular grid high-order finite difference. *Oil Geophysical Prospecting*, 2019.

[14] E. M. Saiki and S. Biringen. Numerical simulation of a cylinder in uniform. 123(36):450–465, 1996.

[15] S. Tang. Lagrange's interpolation polynomial. *Journal of Qingdao University of Science and Technology (Natural Science Edition)*, 1992.

[16] Tsugio and Fukuchi. Numerical calculation of fully-developed laminar flows in arbitrary cross-sections using finite difference method. *AIP Advances*, 1(4):121–133, 2011.

[17] Tsugio and Fukuchi. Numerical stability analysis and rapid algorithm for calculations of fully developed laminar flow through ducts using time-marching method. *AIP Advances*, 3(3):32101–32101, 2013.

[18] Tsugio and Fukuchi. Finite difference method and algebraic polynomial interpolation for numerically solving poissonapos;s equation over arbitrary domains. *Aip Advances*, 2014.

[19] C. F. Zheng and Z. Z. Yan. The method of monte carlo markov chain for solving the poisson equation on irregular domain. *Advances in Applied Mathematics*, 2019.

The union of intervals in constructive mathematics

Chuning Liu[1*], Wenqi Shen[2a]

[1]Shenzhen Middle School, Shenzhen, 518024, China

[2]The Affiliated International School of Shenzhen University, Shenzhen, 518066, China

[a]shenwenqi0601@hotmail.com

*Corresponding Author: Chuning Liu: 2453835242@qq.com

Abstract. Constructive Real Numbers (CRNs) are the concept of real numbers used in Constructive Mathematics, where the objects are thought of as being generated by a computer program. This paper focuses on the property of unions in the constructive end, which means that we study the union of intervals from the viewpoint of Constructive Mathematics. In the end, we prove that for $\mathrm{t < u \leq v < w}$ ($\mathrm{t, u, v, w}$ are CRN), then it is always true that $\mathrm{[t, v] \cup [u, w] \subset [t, w]}$ but generally it is not true that $\mathrm{[t, w] \subset [t, v] \cup [u, w]}$. However, for $\mathrm{t < u < v < w}$($\mathrm{t, u, v, w}$ are CRN), it is always true that $\mathrm{[t, w] = [t, v] \cup [u, w]}$.

1 Introduction

The idea of a real number utilized in constructive mathematics is known as a Constructive Real Number (CRN). It is a real number that can be constructed using a variety of constructive approaches in the broadest sense. In this paper, we study the properties of unions of intervals of Constructive Real Numbers CRN. There are three main research results, Step 0, Main Step, and Step 2. In Step 0, we use the definition of "greater or equal and greater" in CRN to show for $t < u \leq v < w(t, u, v, w$are CRN), it is true that $[t, v] \cup [u, w] \subset [t, w]$. In the Main Step, we use an unextendible program to prove that for $t < u \leq v < w(t, u, v, w$ are CRN), it is generally not true that $[t, w] \subset [u, w] \cup [t, v]$. Then in Step 2, we use Cauchy Sequences equipped with convergence regulators to prove that for $t < u < v < w(t, u, v, w$ are CRN), we always have that $[t, w] = [u, w] \cup [t, v]$.

Remark: Of course, when t, u, v, w are the usual real numbers related by the above inequalities then it is always true that $[t, w] = [u, w] \cup [t, v]$. For obvious reasons, however, as we shall see in the case of t, u, v, w being CRN the problem is interesting and nontrivial.

2 Theory

1) Definition of CRN: CRN is defined to be the word $\alpha\Diamond\beta$. Here α, β are two computer programs: α produces a Cauchy sequence of rational numbers, β called the convergence regulator tells how fast the sequence converging is, see Bishop [1] and Kushner [1,2]. They were first introduced by Turing [3].

2) Definition of the unextendible program: unextendible program means that there exists a computable function whose inputs are positive integers that takes values only 0 and 1 and that does not have an everywhere-defined computable extension, see Theorem 10 in Shen and Vereshchagin's book [4].

3) Definition of Cauchy Sequence: A Cauchy Sequence is a progression of items that are arbitrarily near to one another (as shown in Figure 1). More specifically, all sequence elements except for a finite number are closer together than the given distance for any small positive distance.

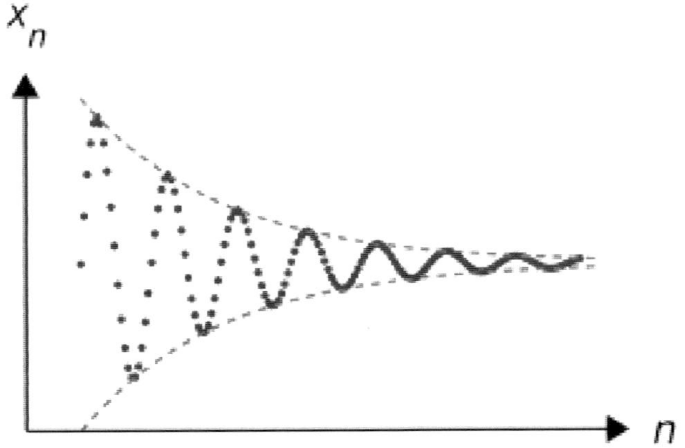

Figure 1: Cauchy Sequence

4) The definition of the standard regulator is that for all N and $i, j \geq N$ we have $|\alpha(i) - \alpha(j)| < 2^{-N}$. Without the loss of generality, we can assume that all CRNs have standard regulators, as seen in Bishop's book.[1]

5) The mathematical study of the properties that are maintained via deformations, twistings, and stretchings of objects is the definition of topology (Figure 2 is a typical example of these kinds of problems).

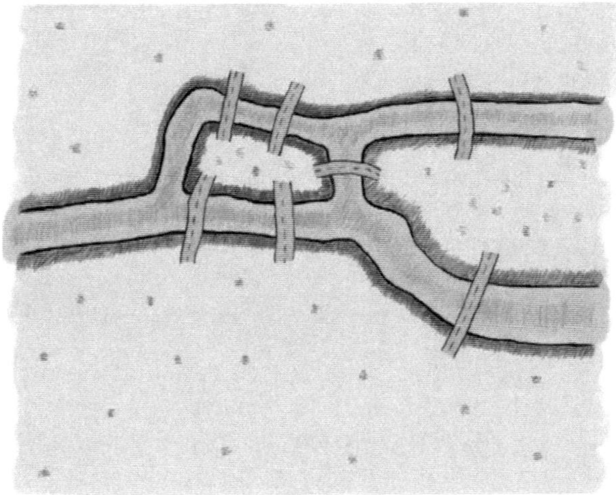

Figure 2: A typical topologic question: Seven Bridges of Königsberg

6) Definition of topological space a set 1(whose elements are called points) and a function (called closure) assigning to each set X⊂ 1 a set $\overline{X}$ ⊂ 1 satisfying the following four axioms:

Axiom 1. $\overline{X \cup Y} = \overline{X} \cup \overline{Y}$.

Axiom 2. X⊂ $\overline{X}$.

Axiom 3. $\overline{0} = 0$.

Axiom 4. $\overline{\overline{X}} = \overline{X}$

If, moreover, the following axiom is satisfied:

Axiom 5. $\overline{(p)} = (p)$ where p∈1

The space is called a T_1-space (1) (or topological in the strict sense in contrast to general topological space satisfying Axioms 1-4).

7) A topology on a set X is defined as a nonempty collection of subsets of X. This collection is referred to as an open set, and it includes both X and the empty set. Any union of open sets and any finite intersection of an open set are also considered to be open sets. A topological space is a set that has a topology on it.

3 Proof

3.1 Step 0

In this section, we show that $[t, v] \, U \, [u, w] \subset [t, w]$, for all t, v, u, w that are constructive real numbers (CRN) and $t < u \leq v < w$. To prove this statement, we need to choose a CRN c in $[t, v]$ and prove $t \leq c \leq w$ then we prove a similar statement for c in $[u, w]$

Proof:

First, we let c be in $[t, v]$, which is defined t $\{c \mid t \leq c \leq v\}$. In this interval, to prove $t \leq c \leq w$, we need to prove $c \leq w$.

According to the assumption, we know that $c \leq v, v < w$.

By the definition of the comparison inequalities for the CRN, $c \leq v$ means that

$\forall \, rational \, r > 0$, $\exists N_1$, such that $\forall n, m \geq N_1, we \; have \; that \; -r \leq v_m - c_n$.

Here and below the words there exists meaning that the object can be found algorithmically.

$v < w$ means that

$\forall \, rational \, r > 0$, $\exists N_2$, such that $\forall \, n, m \geq N_2, we \; have \; that \; -r \leq w_n - v_m$.

We choose $N = Max(N_1, N_2)$, then $\forall n, m \geq N$, we have

$c_n - r \leq v_m$ and $w_n + r \geq v_m$

We get $c_n - r \leq v_m \leq w_n + r$ and then get $c_n - r \leq w_n + r$

Finally, we can conclude that $-2r \leq w_n - c_n$

Since we assume that r is arbitrary, so $c \leq w$

□

The proof of the inclusion $[u, w]$ into $[t, w]$ is similar. We let c be in [u,w], which is defined $\{c \mid u \leq c \leq w\}$. In this interval, to prove $t \leq c \leq w$, we need to prove $t \leq c$.

According to the assumption, we know that $u \leq c, t < u$.

By the definition of the comparison inequalities for the CRN, $u \leq c$ means that

$\forall$ rational $r > 0, \exists N3$, such that $\forall n, m \geq N3$, we have that $-r \leq c_m - u_n$.

Here and below the words there exists meaning that the object can be found algorithmically.

$t < u$ means that

$\forall$rational $r > 0, \exists N_4$, such that $\forall n, m \geq N_4$, we have that $-r \leq u_n - t_m$.

We choose N = Max (N3, N4), then $\forall n, m \geq N$, we have

$$u_n \leq c_m + r \; and \; u_n \geq t_m - r$$

We get $t_m - r \leq u_n \leq c_m + r$ and then get $t_m - r \leq c_m + r$

Finally, we can conclude that $-2r \leq c_m - t_m$

Since we assume that r is arbitrary, so $t \leq c$

In conclusion, because $t \leq c \leq w$, we have $[t, v] \cup [u, w] \subset [t, w]$

3.2 Main Step

In this section, we prove that it is generally not true that $[t, w] \subset [t, v] \cup [u, w]$. In constructive mathematics, the statement that an element belongs to the union of two sets means that you can algorithmically find the set it belongs to. To be specific, our statement means that there does not exist a program that given a CRN $z_k \in [t, w]$ says that $z_k \in [t, v]$ or that $z_k \in [u, w]$. We show this by contradiction in the particular case when $u = v$ which of course implies that the inclusion cannot be true for arbitrary u and v with $u \leq v$.

Proof:

We assume that $[t, w\ ] \subset [t, v] \cup [u, w]$ and let H be an unextendible program from positive integers to 0,1. We define $a\ sequence\ z_k\ of\ CRNs\ via\ z_k(n) = u_m - 2^{-m-s}$ iff H terminates on step m and $n \geq m$ and the result of H is 0 (s is a positive integer such that $u - t > 2^{-s}$ and $w - v > 2^{-s}$ which can be found since $t < u$ and $v < w$ are CRNs).

So, in this case, the CRN Z_k is in $[t, v]$ and not in $[u, w]$.

If H terminates on step m and $n \geq m$ and H produces 1, then we define $z_k(n) = u_m + 2^{-m-s}$, so z_k is in $[u, w]$ and not in $[t, v]$. If we are at the beginning of generating the CRN z_k, or if H does not terminate ever, then $z_k(n) = u_m$. Since $u = v$, both answers $z_k \in [t, v]$ and $z_k \in [u, w]$ are correct.

Let G be a program for every CRN $z_k$in the interval $[t, w]$ tells a correct answer that whether Z_k is in $[t, u]$ or in $[u, w]$ (If $z_k$is in the overlap of the two intervals, then G is allowed to give any one of the two correct answers). If G tells that $z_k(n)$ is in the interval $[t, v]$, then we put $H'(n) = 0$ and if G tells that $z_k(n)$ is in the interval $[u, w]$, then we put $H'(n) = 1$. This H' is clearly the extension of H to all positive integers.

However, our H is an unextendible program (the definition is at the beginning).

Thus, such a program G that (given a CRN Z in $[t, w]$) always tells if Z is in $[t, v]$ or $[u, w]$ does not exist. Thus, from the viewpoint of Constructive Mathematics, it is not true that

$$[\,t, w\,] \subset [t, v] \cup [u, w].$$

After the proof for the main step, we have a further question: does such a program still exist if $t < u = v < w$ are rational numbers? We can get the result from the proof above. We do focus on the fact that $u = v$ in the proof above, but we are not considering the fact that t, u, v, w are actually CRNs instead of their particular case-rational numbers. So, our proof also shows that with four rational numbers $t < u = v < w$, there is no program to detect whether a given CRN Z, which originally is in $[t, w]$, in $[t, u]$, or in $[v, w]$.

3.3 *Step 2*

In the Main Part, we have already proved that for $t < u \leq v < w(t, u, v, w$ are CRN), it is generally not true that $[t, w] \subset [u, w] \cup [t, v]$. After doing some further study, we discover that when the condition change from $u \leq v$ to $u < v$, the proposition $[t, w] \subset [u, w] \cup [t, v]$ is always true. Below, we will show you how we get to this conclusion.

Proof:

Take $t < u < v < w$; t, u, v, w are CRN.

Without the loss of generality, we assume that the Cauchy Sequences defining t, u, v, w have standard regulators. We compute their elements u_n, v_n till

$$v_n - u_n > \frac{4}{2^n}$$

Such a moment will happen because CRN $v > u$

Take any CRN z and compute it till z_k, such that $k > n$

If $z_k < v_n - \frac{1}{2^n}$, decide that $z \in [t, v]$; In all other cases decide that $z \in [u, w]$

In that case, there exists a program that can decide whether such a z is in $[t, v]$ or $[u, w]$, which means $[t, w] \subset [u, w] \cup [t, v]$ from the viewpoint of Constructive Mathematics.

4 Conclusion

With all three steps above, we have proved several relationships between different unions and intervals, which are that $[t, v]\ U\ [u, w] \subset [t, w]$, for all t, v, u, w are constructive real numbers (CRN) and for $t < u \leq v < w$, it is generally not true that $[\,t, w\,] \subset [t, v] \cup [u, w]$, while the inclusion $[t, w] \subset [u, w] \cup [t, v]$ is true when the condition changes from $u \leq v$ to $u < v$. We prove all the statements step by step with the definition of 'greater and equal or less and equal', constructive real number, extension, and unextendible program. In step 0, we construct the proof in two parts, both $t \leq c$ and $c \leq w$. Then, we use an unextendible program to prove

$[t, w] \subset [t, v] \cup [u, w]$ is not true by contradiction to the fact there exists an extendible program. In step 2, using the definition of Cauchy sequence with standard regulators, we prove that $[t, w] \subset [u, w] \cup [t, v]$ when $u < v$.

Acknowledgment
This research was done during the Hybrid program organized by Neoscholor company. Research projects were created by Viktor Chernov and Vladimir Chernov.

Chuning Liu and Wenqi Shen contributed equally to this work and should be considered co-first authors.

Reference

[1] E. Bishop (2012) "Foundations of Constructive Analysis". Ishi Press
[2] B. A. Kushner (1984) "Lectures on Constructive Mathematical Analysis". Translated by E. Mendelson. American Mathematical Society
[3] A. Turing (1937) "On Computable Numbers, with an Application to the Entscheidungsproblem" Proceeding of the London Mathematical Society, Vol S2-42, Issue 1, pages 230-265
[4] A. Shen and N. K. Vereshchagin (2002) "Computable Functions". Translated by V. N. Dubrovskii. American Mathematical Society. Existence of unextendible programs Theorem 10 in Shen book, existence of the enumerable but undecidable sets Theorem 11 in Shen book

Journal of Physics: Conference Series **2455** (2023) 012019 doi:10.1088/1742-6596/2455/1/012019

Modified Cam-clay Model Parameters M of Kunming Red Clay

Xianru Liu, Jianguang Li, Yi Ren ,Kuangjie Zhao

(Civil engineering College, Southwest Forestry University, Kunming 650000, China)

liu20785393@163.com

Abstract: Red clay is widely distributed in Kunming. At present, there are many engineering projects in Kunming, and many engineering projects are built on red clay. Therefore, it is very important to study the mechanical properties of red clay. Kunming red clay was remolded and prepared, and triaxial consolidation drainage tests of remolded saturated clay with the same water content and different dry densities were designed. Through the triaxial consolidation drainage test, the shear strength parameters of Kunming saturated red clay and the modified Cam-clay model parameters are obtained. The experimental results show that the modified Cam-clay model parameters of saturated red clay with the same water content and dry density are basically equal, and the larger the dry density is, the larger the modified Cam-clay model parameter M of saturated red clay is. The test results can provide reference data for engineering design in the Kunming area and modified Cam-clay model simulation in the Kunming area.

1. Introduction

Red clay is mainly distributed in southern China, especially in Guizhou and Yunnan. Red clay in Yunnan is mainly distributed in Kunming, Wenshan, and other places. With the increasing number of engineering projects in Kunming, the engineering problems of red clay have become increasingly prominent. In the fields of highway construction, subway construction, high-rise building construction, etc., the compression characteristics and shear strength index of red clay are all practical problems that need to be studied. Red clay is a regional special soil that is formed by limestone, basalt[1], shale, and other rocks through intense lateralization, covering carbonate rock series and is brownish red or brownish yellow[2]. It has the characteristics of high liquid limit, high plasticity, large void ratio, and low compactness. Among them, the liquid limit greater than or equal to 50% is primarily red clay, and the liquid limit greater than 45% is secondary red clay[3]. The research on the shear strength of soil by soil mechanics is mostly conducted with remolded soil, which is divided into unsaturated soil and saturated soil, and the research on saturated soil is more extensive[4]. There are many elastic-plastic models for soil. In 1958, Roscoe et al. of Cambridge University in England first put forward the concepts of state boundary and the critical state line. Then, using the hat yield surface, the associated flow law, and the isotropic hardening law, the plastic strain was taken as the hardening parameter. Based on the energy principle, the Cambridge model was established by combining the conventional triaxial test results of normally consolidated and weakly over-consolidated soil. In 1965, Burland adopted different energy equations to get the modified Cambridge model[5-6]. The modified Cam-clay model is a commonly used elastic-plastic model to study clay, and the engineering practice in many areas in China also proves that the modified Cam-clay model is an ideal model for soft clay. At present, clay is widely studied at home and abroad. Liu Luzhi[7] and Others have determined the compression index and rebound index obtained from the compression test and the values of the modified Cam-clay model parameters of Guangzhou soft soil through the compression test of the original soft soil in Guangzhou. Li Peng[8]et al. obtained the relevant parameters of the modified Cambridge model through triaxial tests. Hong Huihuang[9]obtained the modified Cambridge model parameters of Fuzhou soft soil through the indoor test, and simulated and compared the test with finite element software. The calculated values of the finite element are almost the same as the experimental values, which shows the rationality of the parameters obtained from the test.

In this paper, Kunming red clay was selected to prepare test soil samples according to different dry densities. The physical properties of red clay are obtained by conventional physical tests, and the shear strength parameters of red clay are obtained by triaxial consolidation drainage shear test. According to the obtained cohesion and internal friction angle parameters, the failure stress ratio of saturated red clay in the Kunming area is calculated by using the modified Cambridge model theory. The influence of different dry densities on the shear strength index of red clay and the modified Cambridge model parameter M is analyzed, which can provide a reference for Kunming construction engineering practice and geotechnical numerical simulation.

2. Experimental method

2.1 sample preparation

The soil samples used in the test are classified as reddish-brown clay according to GB 0021-2001. The sampling place is the supporting municipal construction site of the K12 school in Kunming Airport Economic Zone, and the sampling depth is 5.8~6.8m. To study the mechanical properties of saturated red clay in Kunming under different dry densities, the red clay was remolded, and the preparation and saturation of samples were strictly in accordance with the requirements of the Geotechnical Test Code[10].

Through drying, weighing, and calculating, it is found that the water content of the original red clay is 30%. Therefore, under the same water content, two kinds of remolded red clay were prepared according to different dry densities. Among them, two groups of soil samples with a water content of 30% and a dry density of 1.45g/cm^3 were prepared, and two groups of soil samples with a water content of 30% and a dry density of 1.5 g/cm^3 were prepared, with a group of four, totaling 16 soil samples. At the same time, the physical properties of Kunming saturated red clay are obtained by routine physical tests such as mineral composition, liquid limit, and plastic limit, as shown in Table 1 below.

Table 1 Physical Properties of Kunming Red Clay

Number	Moisture content %	Dry density g/cm^3	Void ratio	Liquid limit	Plastic limit
1	30.0	1.50	0.933	53.5	36.3
2	30.0	1.50	0.933	53.9	36.1
3	30.0	1.45	1.005	54.8	37.1
4	30.0	1.45	1.005	54.8	36.3

2.2 Test process

In this paper, the strain-controlled triaxial apparatus of Nanjing Soil Instrument Factory, TSZ-3, is used. Two groups of remolded red clay with different dry densities were tested by triaxial consolidation drainage. The sample was made into a standard triaxial sample with a diameter of 3.91cm and a height of 8cm, as shown in Figure 1 below. Then, a vacuum saturates the soil sample and loads it into a triaxial pressure chamber for a triaxial drainage test. According to the order of saturation, consolidation, and shear of clay samples, the test process is shown in Figure 2. During the shearing process, the readings of the force ring gauge and the pore pressure gauge were recorded. After the readings don't increase, we continue to shear by 3%~5% to ensure the critical state. When the dynamometer reading has no peak value, shear until the axial strain is 15%~20%. Each sample is subjected to different confining pressures. Four confining pressures of 100KPa, 200KPa, 300KPa and 400KPa are used in this paper.

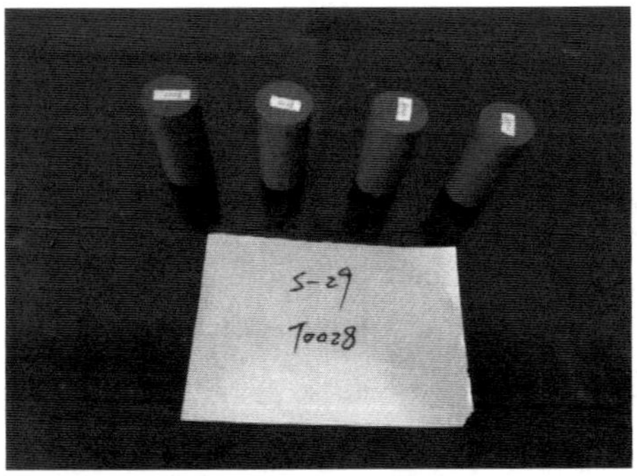

Figure 1 Remodeled specimen

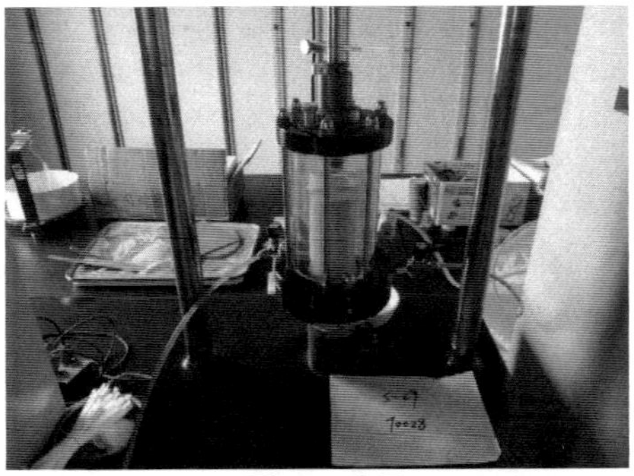

Figure 2 Test process

2.3 test results

According to the test data obtained from the consolidation drainage triaxial shear test, the relationship curve between principal stress difference and axial strain is drawn with principal stress difference as ordinate and axial strain as abscissa, as shown in Figure 3.

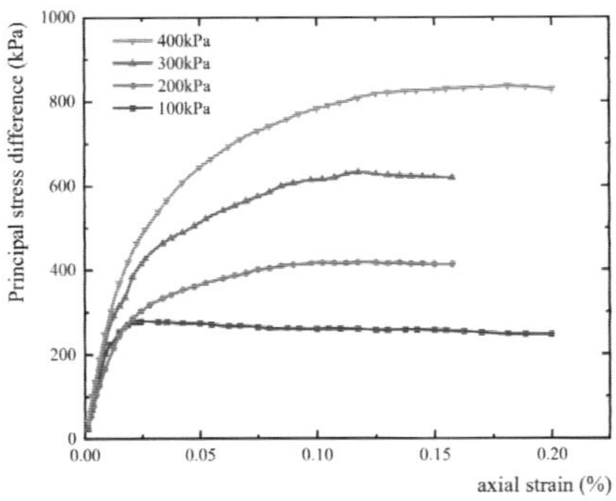

(a) No.1 soil sample

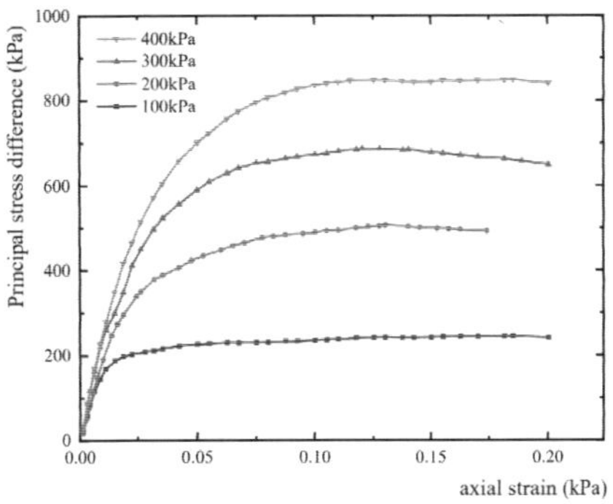

(b) No.2 soil sample

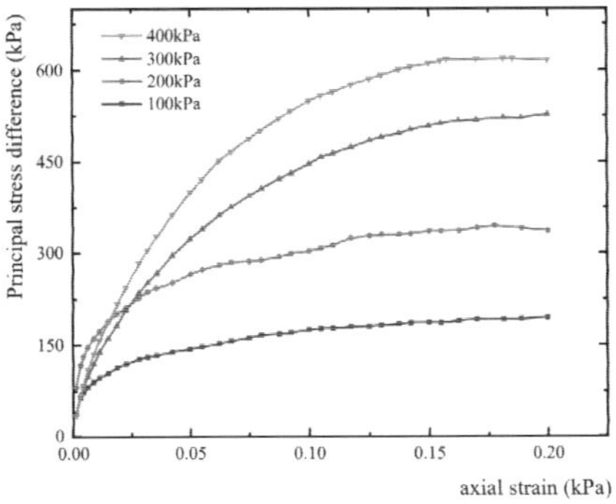

(c) No.3 soil sample

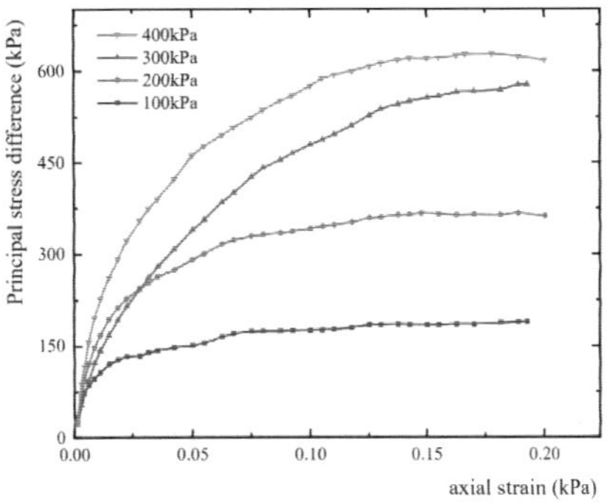

(d) No.4 soil sample

Figure 3 Relationship between principal stress difference and axial strain

Soil samples No.1 and No.2 have a water content of 30% and a dry density of 1.5g/cm^3. And No.3 and No.4 soil samples have a water content of 30% and a dry density of 1.45 g/cm^3. It can be seen from Figure 3 that under different confining pressures, the principal stress difference increases with the increase of axial strain, and the trend of the relationship curve between the principal stress difference and axial strain of the same dry-density soil sample is basically the same.

Taking the peak value of the principal stress difference as the failure point, the damage stress circle and the envelope of the damage stress circle under different confining pressures are drawn, and the total stress intensity parameters are obtained, as shown in Figure 4 below.

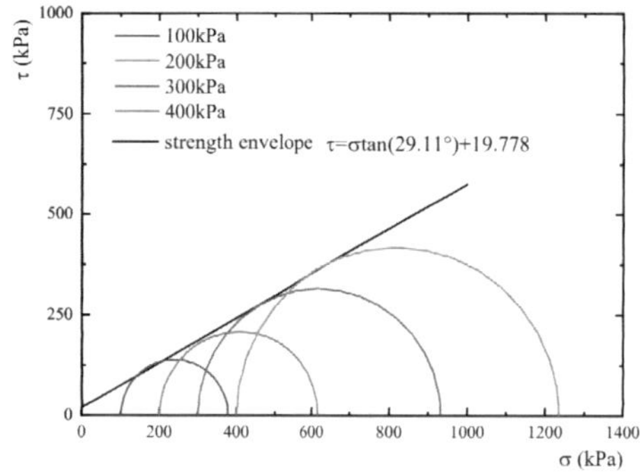

(a) No.1 soil sample

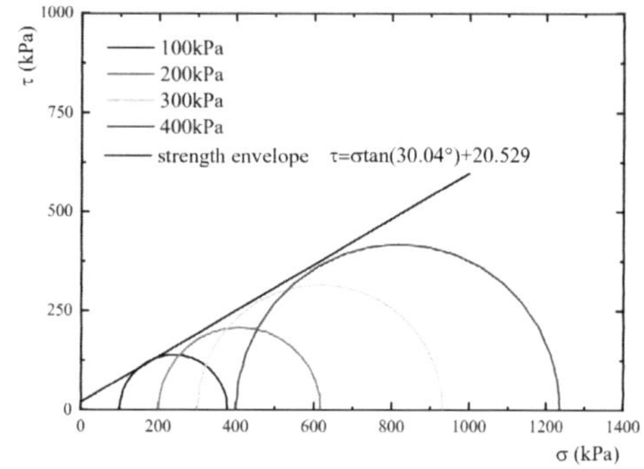

(b) No.2 soil sample

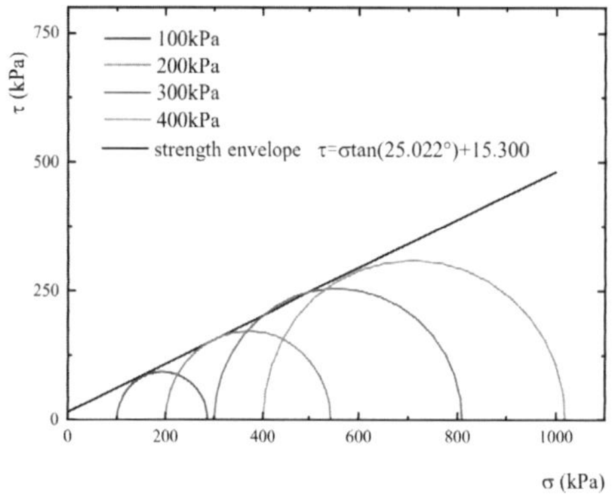

(c) No.3 soil sample

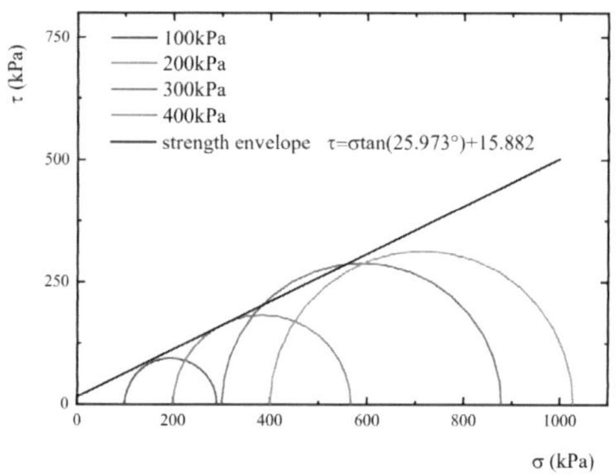

(d) No.4 soil sample

Figure 4 Strength envelope of consolidated drainage shear

According to the obtained strength envelope, the shear strength index cohesion C_{cu} and internal friction angle $\emptyset_{cu}$ of Kunming red clay in the triaxial consolidation drainage test is obtained. The failure stress ratio of red clay under two different dry densities is obtained by formula $M=\frac{6sin\emptyset}{3-sin\emptyset}$. The shear strength parameters of saturated red clay in Kunming under two different dry densities and the calculated values of modified Cam-clay model parameter M are shown in Table 2 below.

Table 2 Shear strength parameters and modified Cambridge model parameters

number	cohesive strength c_{cu}/kPa	Internal friction angle ϕ/°	Critical state stress ratio M
1	19.778	29.106	1.161
2	20.529	30.038	1.202
3	15.300	25.022	0.985
4	15.882	25.973	1.026

According to the internal friction angle $M = \frac{6\sin\phi}{3-\sin\phi}$ and formula calculation, the failure stress ratios of No.1, No.2, No.3, and No.4 soil samples are 1.161, 1.202, 0.985, and 1.026 respectively. The water content of the No.1 soil sample and No.2 soil sample is 30%, and the dry density is 1.5g/cm3. Soil samples No.3 and No.4 have a water content of 30% and a dry density of 1.45g/cm3. It can be seen from Table 1 that the values of the failure stress ratio M of soil samples with the same moisture content and dry density are very close, so the average values are 1.18 and 1.01 for the convenience of the subsequent calculation of the modified Cam-clay model. Under the same moisture content, the greater the dry density of the soil sample is, the greater the value of its failure stress ratio M.

3 Conclusion

Through triaxial consolidation undrained test of saturated red clay in the Kunming area, the shear strength indexes of soil samples were obtained. Its indicators include the internal friction angle and cohesion of soil samples. The required parameter M of the modified Cambridge model is obtained by calculation. At the same time, the failure points of four groups of soil samples under different confining pressures are fitted, and the critical state line of the modified Cambridge model and the fitted failure stress ratio M are obtained. By comparing and modifying the parameter M of the Cambridge model, the following conclusions are summarized:

(1) The values of modified Cambridge model parameters obtained by the internal friction angle of saturated red clay in the Kunming area are 1.18 and 1.01. Among them, 1.18 is the failure stress ratio of red clay with 30% moisture content and 1.5g/cm^3 dry density, and 1.01 is the failure stress ratio of red clay with 30% moisture content and 1.45g/cm^3 dry density.

(2) Under the same moisture content, the dry density has a great influence on the shear strength of saturated red clay in Kunming. The larger the dry density is, the larger the modified Cambridge model parameter M is.

(3) The saturated red clay in Kunming is representative of this paper. However, if we want to comprehensively evaluate the mechanical properties of red clay in Kunming and obtain the parameters of the modified Cambridge model in Kunming, we should study more samples from other areas. Therefore, in the next step, we can conduct a comparative study on the soil samples from different areas in Kunming.

References:

[1] Fu Bichang, Huang Ying, Fang Liping, et al. Cause study of laterite overlying carbonate rocks [J]. Journal of Rock Mechanics and Engineering, 2013, 32(A01):9.

[2] Luo Congcong, Gong Aimin, Xu Xingqian, et al. Comparative Analysis of Shear Characteristics of Undisturbed and Remolded Red Clay in Kunming[J]. Water Power, 2022(002):048.

[3] People's Republic of China (PRC) Ministry of Water Resources. Classification Standard of Soil [M]. China Planning Press, 1991.

[4] Zhao Rui, Zuo Shuangying, Wang Song, et al. Experimental and mechanism analysis of water contents on triaxial Shear Strength of Remodeled Red Clay of Guiyang [J]. Hydrogeology And Engineering Geology, 2015(5):6

[5] Li Guangxin. Advanced Soil Mechanics [M]. Tsinghua University Publishing House, 2004.

[6] Luo Ting. Constitutive relationship of soil [M]. People's Communications Publishing House, 2010.

[7] Liu Luzhi, Yang Tao, Ma Pengzhen. Compressive characteristics and modified Cam-clay model parameters λ and κ of Guangzhou soft soil [J]. Journal of Water Resources and Architectural Engineering, 2016, 14(1):4.

[8] Li Peng, WU GANG, Liu Qiang, et al. Experimental study on parameters of modified Cambridge model of marine clay [J]. Geotechnical Investigation and Surveying, 2020, 48(10):6.

[9] Hong Huihuang. Modified Cam-clay model test and numerical simulation verification of soft soil in Fuzhou area [J]. Fujian Architecture and Construction, 2011(12):3.

[10] Institute of Highway Science, Ministry of Communications. Highway Geotechnical Test Code:

JTG E40-2007[M]. People's Communications Publishing House, 2007.

Journal of Physics: Conference Series **2455** (2023) 012020 doi:10.1088/1742-6596/2455/1/012020

Application of Modular Maximum of Wavelet Transform in the Paint-coat Damage Detection

Faliang Sun*, **Xiaohuan Guo, Xiangyan Luo and Xiaoju Yong**

Sergeants School of Aviation Lecomotive, Air Force Engineering University, Henan

434506563@qq.com

Abstract: The reflected waves of each interface might obliterate each other and be indistinguishable in the conventional ultrasonic testing of the multi-interface heterogenous thin-layer structure. Aiming at this problem, the method of dealing with signals by the modular maximum of wavelet transform is put forward and the detection testing for the embedded defects of test pieces is implemented. The test results show that this method can judge whether there is a defect by identifying the time point of $\overline{|W_{m_1} f(t)|}$.

1 Introduction

Currently, the application of all kinds of coating layers and surface-modified layers with a thickness scale of micrometer or even nanometer is very wide. It has important theoretical and practical value to implement nondestructive tests for their defects. It is a hot spot of the modified layer quality assessment to characterize the surface-modified layers' characteristics (such as the thickness, elastic modular, bonding strength and residual stress) of the material by the method of ultrasonic testing. Ultrasonic reflected wave is a comprehensive response of sound wave, system and medium, containing a lot of useful information such as the nature of internal defects, the state of material structure, and the state of stress distribution. The goal of nondestructive characterization can be achieved by analyzing the acquired signals of reflected waves. However, the thicknesses of thin-layer materials are small (they are mainly in the scale of a micrometer or nanometer) and they are usually used together with adhesive layers and the matrix to form a three-layer or even multi-layer structure, bringing great difficulty to the acquisition and the analysis of ultrasonic signals. The mechanical and physical performance data such as the thickness of layers, elastic modular and Poisson's ratio have no uniform standard, which also brings difficulty to the layer assessment [1].

Based on the Fourier transform, the wavelet transform is a new analysis method in recent years. This method breakthroughs the time domain limit of not having the resolving power of the conventional Fourier transform [2]. It can analyze signals in a specific time interval and frequency range. Besides, this method can well analyze the localization of the frequency range and the time interval. At the same time, it can focus on every detail in the signal because the sampling step of the time interval applied in its frequency is in the form of gradual refinement. However, to express the three-dimensional information of the time, frequency and amplitude, in general circumstances, the time-scale figure is used to show the result of wavelet analysis [3]. Although the time-scale figure is directly perceived through the senses and easy to understand, its temporal location is bad which can't satisfy the requirements for real-time nature in the project [4]. Therefore, this paper established an ultrasonic testing model for the stealthy coating. A method processing the ultrasonic signals of the modular maximum of wavelet transform was put forward based on the ultrasonic reflected wave signals to achieve the damage detection of the layers of stealthy layers.

2 Detection Principles

2.1 Wavelet transform principle

The wavelet transform is a special waveform with limited length and an average value of 0, which calculates the inner product of the wavelet function and original signal at different scales after a certain primary function is shifted [5]. From the point of view of mathematics, the function $\psi(t)$ of the mother wavelet shall satisfy two conditions:

(1) $\psi(t)$ square integrable, that is $\psi(t) \in L^2$ (R)

(2) $\psi(t)$ is the Fourier transform of the function $\psi(t)$, $\int_{-\infty}^{+\infty} |\hat{\psi}(t)|^2 |\omega|^{-1} d\omega < +\infty$

After a mother wavelet is calculated, it can be translated and expanded and contracted to obtain:

$$\psi_{a,b}(t) = \frac{1}{\sqrt{a}} \psi\left(\frac{t-b}{a}\right), (a > 0, b \in R), \quad (1)$$

Then $\psi_{a,b}(t)$ is called the wavelet primary function. In the formula (1.1): a is called the scale factor and b is called the scale translation factor. At the same time, if a and b are both continuous, then $\psi_{a,b}(t)$ is called continuous wavelet primary function. In this case, let $\psi(t)$ be a mother wavelet while $\psi_{a,b}(t)$ is a continuous wavelet primary function acquired by expanding and contracting and translating. Then, for any function $x(t)$, its wavelet transform is:

$$WT_x(a,b) = \frac{1}{\sqrt{a}} \int_{-\infty}^{+\infty} X(t) \overline{\psi}\left(\frac{t-b}{a}\right) dt \quad (2)$$

Then WT_x (a,b) is called the continuous wavelet transform (that is CWT) of x(t). Specifically, $\overline{\psi}(t)$ is the conjugation of $\psi(t)$.

According to the definition of the continuous wavelet transform, when the scale factor a and the scale translation factor b is transforming continuously, the information of the continuous wavelet transform of the signal x(t) is redundant. From the point of view of easy calculation and economical calculation amount, the authors hope to reduce the redundancy of wavelet transform coefficients as much as possible without losing the original signal x(t) information to reduce the calculation amount. In the practical application, the continuous wavelet transform factors a and b are usually discretized. The specific process is: The scale factor a and scale translation factor b in the mother wavelet function $\psi_{a,b}(t)$ are limited to fixed discrete points to have short-cut processes.

For the scale factor a, it is usually discretized according to power series, that is let a be a^0, a^1.... a^j. Then let $\psi_{(a,b)}(t)$ be discretized as:

$$a_0^{\frac{j}{2}} \psi[a_0^{-j}(t-b)], j = 0,1,2 \ldots \quad (3)$$

For the scale translation factor b, in order to not lose information, the sampling must satisfy the Nyquist Sampling Theorem, that is, the sampling frequency is not less than twice the frequency passband of the scale. That is, let the sampling interval of the shift amount b under a specific scale j be $\nabla b = a_0^j b_0$. In this case, $\psi_{(a,b)}(t)$ is transformed to:

$$a_0^{\frac{-j}{2}} \psi[a_0^{-j}(t - k a_0^j b_0)], j = 0,1,2 \ldots, k \in Z \quad (4)$$

However, in the practical application, in order to calculate easily, it usually lets $a_0 = 2$, $b_0 = 1$. In this case, $\psi_{(a,b)}(t)$ is:

$$\psi_{(j,k)}(t) = 2^{\frac{-j}{2}} \psi(2^{-j}t - k), j = 0,1,2 \ldots, k \in Z \quad (5)$$

Specifically, $\psi_{j,k}(t)$ is called the discrete wavelet basis, then the expression of the discrete wavelet transform of the function x(t) is:

$$WT_x(j,k) = \int X(t)\,\psi_{j,k}^{*} dt \tag{6}$$

In this case, the discrete wavelet transform will not cause the information loss of the original signals.

In conclusion, the function after the wavelet transform is two-dimensional [6]. That is, the wavelet transforms the original one-dimensional signal into a two-dimensional signal, achieving the analysis of signals in the time and frequency domain. In essence, the wavelet transform represents any function x(t) in L^2(R) space as a projection on $\psi_{a,b}(t)$ with different scale translation factors and scale factors. Through adjusting the scale factors and the scale translation factors, the wavelet transforms with different time-frequency widths can be acquired and they can match any time point in the original signal to achieve the local two-dimensional analysis of the time-frequency of the signal.

2.2 The mutation principle of detection signals based on the modular maximum of the wavelet transform

In the ultrasonic testing of the thin-layer material, the reflected waves of the two interfaces between the thin-layer, the medium and the matrix are overlapped, and the signals are distorted, resulting in mutation points. The signal mutation is often the key to signals to be analyzed [7].

When $\psi(t)$ is the nth reciprocal of a specific smooth function $\theta(t)$, this smooth function satisfies

$$\int_{-\infty}^{+\infty} \theta(t)\,dt = 1 \; and \; \theta(t) = 0\big(1/(1+t^2)\big) \tag{7}$$

In this case, $\psi(t)$ satisfy the restraint condition.

We define $\psi^1(t)$ and $\psi^2(t)$ as the following wavelet transforms:

$$\psi^1(t) = d\theta(t)/dt, \psi^2(t) = d^2\theta(t)/dt^2 \tag{8}$$

We record θ_m=$\theta(t/s)/s$. Then for a real function $f(t) \in L^2$(R), there is the following wavelet transform

$$\begin{gathered} W_m^1 f(t) = W^1 f(s,t) = f * \psi_m^1(t) = f * (s \cdot d\theta_m(t)/dt) = sd\big(f * \theta_m(t)\big)/dt = \\ W_m^2 f(t) = W^2 f(s,t) = f * \psi_m^2(t) \end{gathered} \tag{9}$$

It can be seen that $W_m^1 f(t)$ and $W_m^2 f(t)$ are the first derivative and second derivative of the function directly proportional to $\theta_m(t)$ after polishing [8]. According to the mathematical analysis, the extreme point of the first derivative of the function corresponds to the zero point of its second derivative. At the same time, it is also the inflection point of the function itself. The maximum value of the absolute value of the first derivative corresponds to the mutation of the function, while the minimum value (not equal to 0) corresponds to the gradual transition of the function. Therefore, the maximum amplitude point of $W_m^1 f(t)$ corresponds to the mutation point of $f(t)$, and the zero point of $W_m^2 f(t)$ is consistent with the inflection point of $f*\theta(t)$.

Therefore, if the wavelet is chosen as the first derivative of the smooth function, then the mutation point of signal $f(t)$ can be detected through the maximum amplitude point of the wavelet transform $W_m^1 f(t)$ [9]. This is the basic principle of the wavelet transform's application in signal mutation detection. When the ultrasonic wave transmits in the thin-layer structure, the reflected wave signals are distorted due to the overlay of the reflected waves of the heterogeneous interfaces. The method of the modular maximum of the wavelet transform can position the occurrence of distorted signals to achieve the goal of detecting defects [10].

3 Testing

3.1 Sample preparation

The test sample was made by spraying ZrO_2 coating on the titanium alloy matrix with the plasma spraying method. The matrix was T300 and the thickness was about 5 mm. The coating consisted of a primer layer, intermediate layer and surface layer. The thickness of the primer layer was about 50 μm, the thickness of the surface layer was about 30 μm and the thickness of the intermediate layer was about 460 μm. The total thickness of the coating was about 540 μm and the size was 180 * 180 mm. Defects were made by spraying varnish after digging pits, and circular defects (about 75 μm deep) were embedded previously in the matrix (titanium alloy) and the primer layer. The embedded defects were distributed on the sample. The surface appearance of the ZrO_2 coating layer and titanium alloy matrix of the sample was shown in Figure 1.

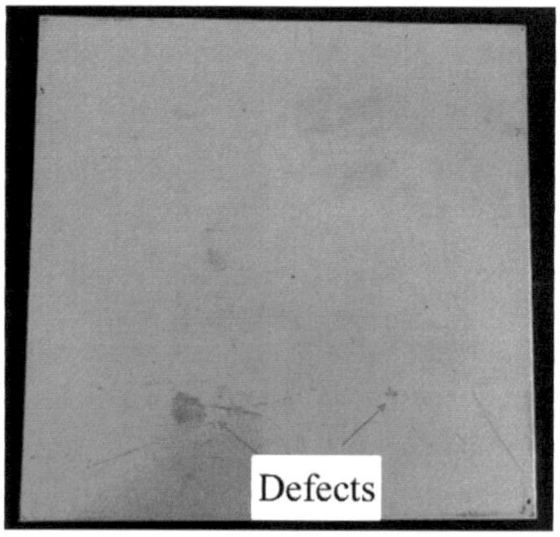

(a) ZrO_2 coating surface

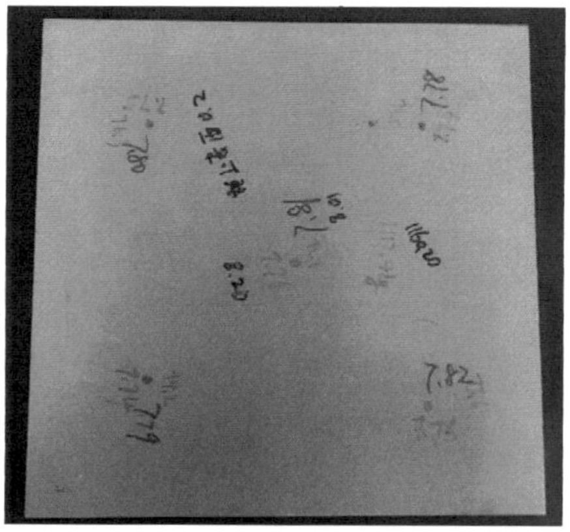

(b) Titanium alloy matrix surface

Figure 1 Coating Test Sample

3.2 Experimental system and parameters

The coating thickness usually ranges from tens to hundreds of micrometers. For the ultrasonic testing for the coating sample, if the direct contact method is used, laying the probe on the coating surface, then one reflected wave of the coating layer is difficult to separate from the initial pulse due to the Coulomb Blockade of the ultrasonic instrument. The result isn't good for analysis. Therefore, the method of setting a delay line between the probe and the sample surface is usually adopted. In this experiment, the probe was placed in the matrix layer used to set the delay line, which can also achieve good detection results.

In this experiment, the direct contact method of the straight probe of the ultrasonic longitudinal wave was used for the detection. The way of receiving signals was self-sending and self-receiving. The experimental parameters were as follows: The probe type was KGKB5C6I. The crystal plate diameter was 5 mm. The central frequency was 10 Mhz. The type of pulse-transmitting receiver was JPR600C. The sampling frequency was 100 Mhz. The experimental system of ultrasonic testing is shown in Figure 2.

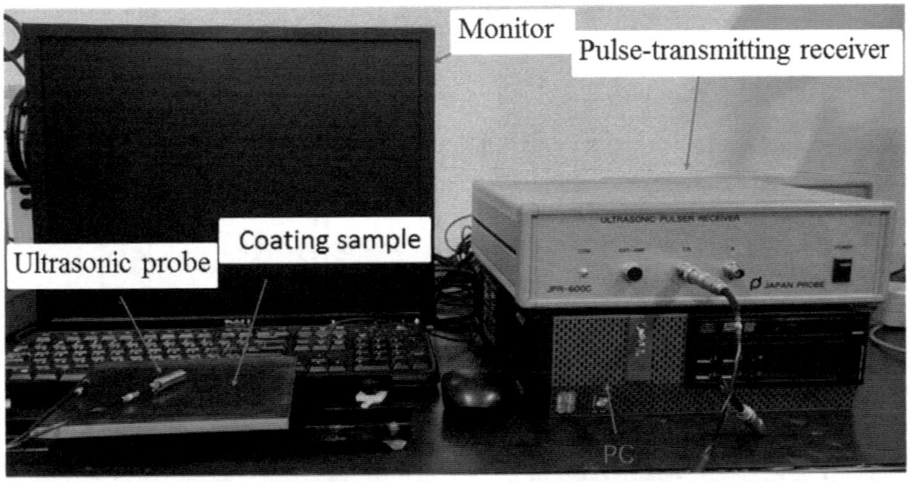

(a) The experimental system of the ultrasonic testing

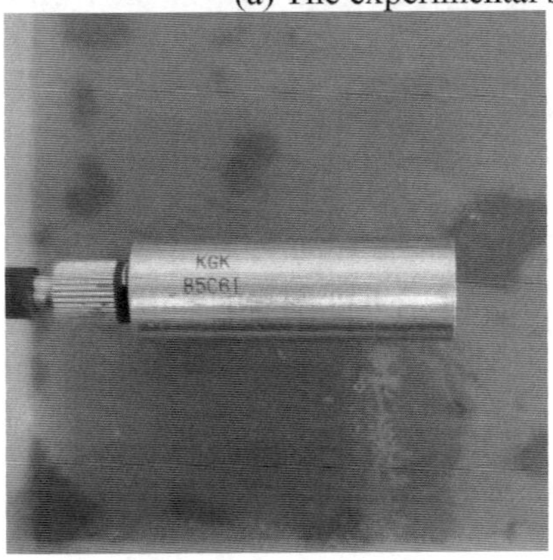

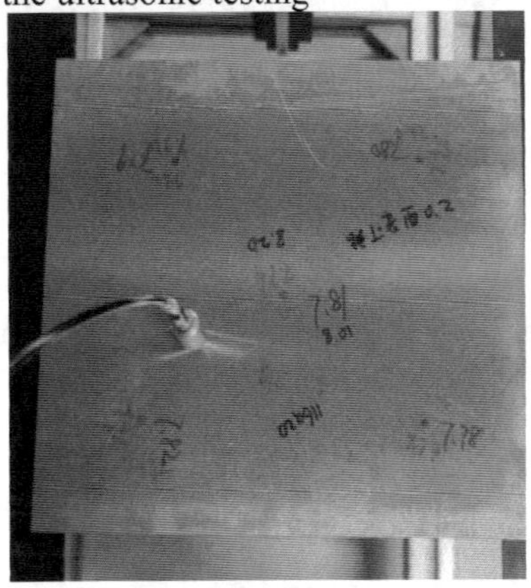

(b) Ultrasonic probe (c) Layout of experimental operation

Figure 2 Ultrasonic Testing System and Experimental Layout

4 Experimental Result and Analysis

4.1 The complete coating layer

To ensure the reliability and accuracy of the experimental data, under the condition of the same experiment and instrumental parameters, the data was collected by placing the probe randomly at 10 defect-free points of the sample. Figure 3 is the normalized time-domain oscillogram of a defect-free point of the sample collected.

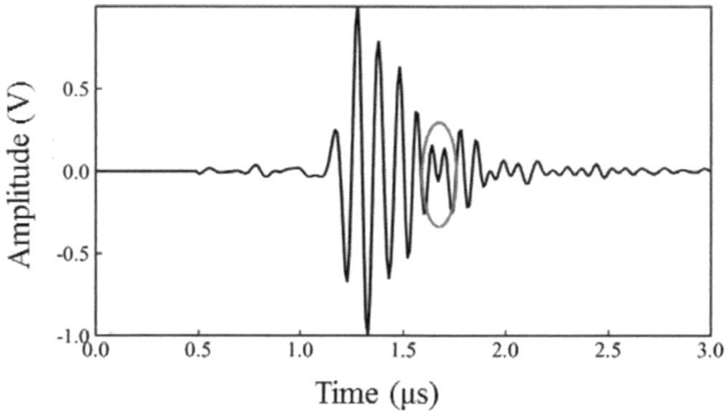

Figure 3 Normalized Time-domain Oscillogram of the Ultrasonic Signal

As shown in Figure 3, in the ZrO_2 coating layer, the overlay of reflected waves occurred in two interfaces when the ultrasonic wave was transmitting. One interface is between the ZrO_2 coating layer and the titanium alloy matrix, and the other interface is between the ZrO_2 coating layer and the air. The position marked with an ellipse in the figure was the position generated by the overlay of reflected waves. At this position, there is a mutation of waveform and it is impossible to directly judge the specific position of the reflected wave generated on these two interfaces in the time domain. Therefore, this experiment implemented the continuous wavelet transform of the original signal in the time domain through Python software and acquired its modular maximum. Taking the sampling point as an example, Figure 4 is the local modular maximum figure of the wavelet transform obtained after the wavelet transform was processed through the Cgau8 wavelet. If the signal had singularity at some time, the connection line of the modular maxima in Figure 4 would change on a large scale. To reduce the error caused by using a single scale to determine singular points, all scales of each sampling point were calculated to acquire the average value of the modular maximum to obtain the average amplitude representing the singularity degree of each point. The average amplitude of the modular maximum of this signal is shown in Figure 5.

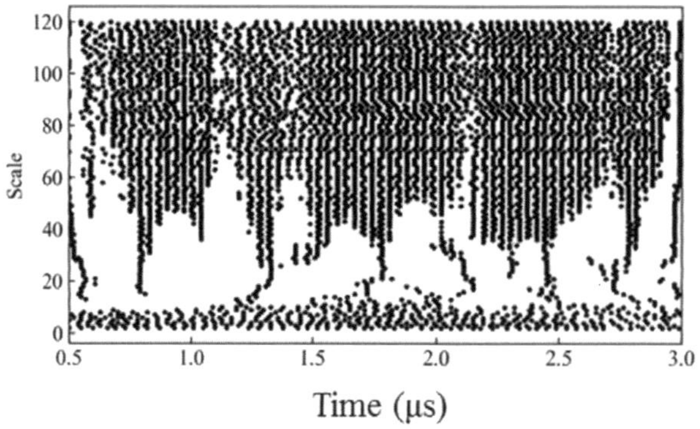

Figure 4 Local Modular Maximum Figure of the Wavelet Transform

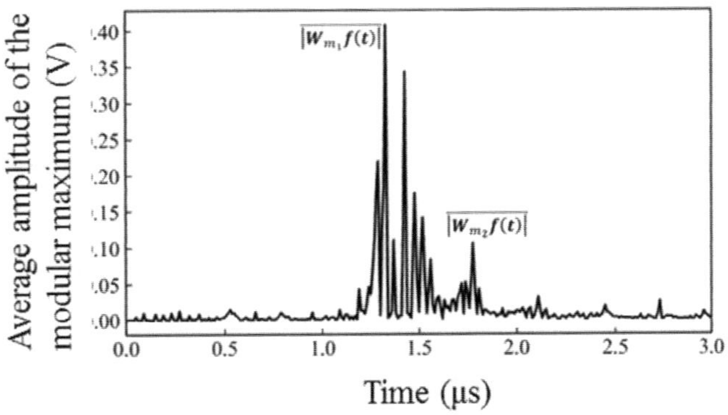

Figure 5 Average amplitude $\overline{|W_m f(t)|}$ of the modular maximum of the wavelet transform

According to the detection principle of the signal mutation, at the interface of the titanium alloy matrix/ZrO_2 ceramic coating and ZrO_2 ceramic coating/air, the energy of $\overline{|\boldsymbol{W_m f(t)}|}$ was the largest. Besides, according to Figure 4 and Figure 5, $\overline{|\boldsymbol{W_{m_1} f(t)}|}$ and $\overline{|\boldsymbol{W_{m_2} f(t)}|}$ corresponded to the titanium alloy matrix/ZrO_2 ceramic coating and ZrO_2 ceramic coating/air interface respectively, therefore the time point data corresponding to the titanium alloy matrix/ZrO_2 ceramic coating and ZrO_2 ceramic coating/air interface were obtained. In addition, the time for the ultrasonic waves to travel back and forth in ZrO_2 ceramic coating was the difference between $\overline{|\boldsymbol{W_{m_1} f(t)}|}$ and $\overline{|\boldsymbol{W_{m_2} f(t)}|}$. Let the longitudinal wave velocity of the coating be 3,002.4 m/s, the coating thickness could be calculated according to the ultrasonic thickness measurement formula.

To discuss the practicability and accuracy of this method, through the method of calculating average value by multiple times of sampling, the coating thickness obtained in the experiment was compared with the actual coating thickness to alleviate the operational error caused by the single experiment. Its result was shown in Table 1.

Table 1 The Coating Thicknesses of 10 Different Sampling Point Positions

Number of the sampling point positions	Calculated thickness/μm	Relative error
1	735.59	36.22%
2	600.48	11.2%
3	630.5	16.76%
4	555.44	2.86%
5	555.44	2.86%
6	555.44	2.86%
7	480.38	11.04%
8	555.44	2.86%
9	660.53	22.32%
10	720.58	33.44%

According to Table 1, the relative errors of point positions 1, 9 and 10 were rather great, seriously deviating from the actual coating thickness. They were excluded considering that it may be caused by non-uniform spraying of coating thickness and operational error during sampling. In this case, the average coating thickness of the other 7 point position was 561.87 μm and the average relative error was 4.05%. It can be seen that under the condition that the ultrasonic wave in the thin layer medium would have the phenomenon of reflected waves overlay at the multi-layer interface position, it was feasible to use the multi-resolution analysis ability of wavelet transform and the "zooming" characteristics to select the Cgau8 wavelet basis and apply the modular maximum of the wavelet transform to the ultrasonic testing of ZrO_2 coating thickness.

4.2 Defected coating

Under the condition of the same experiment and instrument parameters, the probe was placed in the embedded defects of the sampling piece to collect data 3-5 times per position to ensure the repeatability and accuracy of the experiment. Figure 6 is the normalized time-domain oscillogram and the time-frequency figure of the wavelet transform of the embedded defects of the sampling piece collected in the experiment.

By positioning the time in the calculation, it could be calculated that the coating thickness of the embedded defects was about 630.5 μm which was the same as the thickness of ZrO_2 coating plus varnish. It was illustrated that the defects could be identified and positioned through the occurring time position of the modular maximum of the wavelet transform of the embedded defects. Therefore, the measurement of the coating thickness of the defect could be achieved.

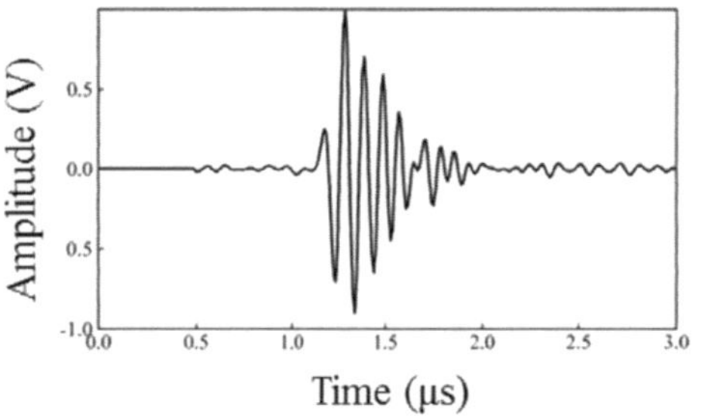

(a) Normalized time-domain oscillogram of the ultrasonic signals of the embedded defects

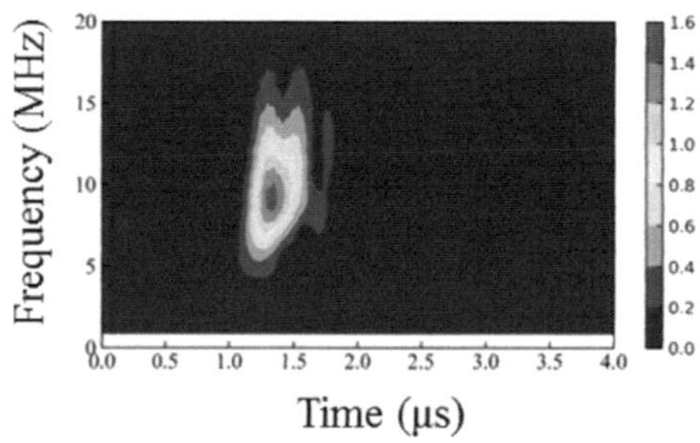

(b) Time-frequency figure of the wavelet transform of the embedded defects

Figure 6 Ultrasonic Signals of the Embedded Defects

5 Conclusion

According to the experiment, the method of the modular maximum of the wavelet transform based on the Cgau8 wavelet applies to the detection of mutation signals. The results of multiple positions on the ZrO_2 coating were averaged and the average relative error was 4.05%. It shows that this method can be adopted in the detection of singularities generated on different interfaces by the ultrasonic signals of the ZrO_2 coating with a thickness of about 540 μm, which has significant meaning to the thickness measurement of the coating. At the same time, this method is used to experiment and analyze the embedded defects of the sampling pieces. The result shows that this method can judge whether there is a defect by identifying the $|W_{m_1}f(t)|$ time point illustrating that the occurring time point of the modular maximum of the wavelet transform of the embedded defects can be adopted to identify and position the defects.

References

[1] Huang H. T. Research on the Processing Method of Ultrasonic Signal Based on the Time-frequency Analysis [D]. Yangtze University, 2020.
[2] Sun yunlian, Liu dunmin. Time-frequency analysis and wavelet transform and its application [J]. Wuhan University Journal (EngineeringEdition), 2003(02): 103-106.
[3] Yang xudong. ECG signal feature parameter Extraction based on wavelet transform [D]. University of Electronic Science and Technology of China,2020.

[4] Nie Y. Research on the Processing Method of Ultrasonic Signal Based on the Time-frequency Analysis [D]. Dalian University of Technology, 2006.
[5] Yu kaiping, Zou jingxiang, Yang bingyuan. Study on the properties and applications of wavelet function [J]. Journal of Harbin Engineering University, 2000, 032(002):36.
[6] Morlet J, Arens G, Fourgeau E, et al. Wave propagation and sampling theory-Part I: Complex signal and scattering in multilayered media [J]. Geophysics, 2013, 47(2):203.
[7] Wang X. G., Chang J. J., Shan Y. C., Xu J. J., Yao M. Ultrasonic Testing of Thickness of the Thin Rubber Layer Based on the Modular Maximum of the Wavelet Transform [J]. Journal of Dalian Maritime University, 2007(04): 126-129.
[8] Du jingjing, Zhang junfang, Liu jianxiao, et al. Detection of coating signal singularity based on wavelet transform mode maximum method [J]. Technology Information, 2022, 20(05): 16-18.
[9] Zhan huajun, Zhu jian. Research on Crack identification and location of portal frame based on mexh wavelet transform [J]. Shanxi Architecture, 2016, 42(24): 43-45.
[10] Wang xinngguo, Chang junjie, Shan yingchun, et al. Ultrasonic measurement of thin rubber layer thickness based on wavelet transform mode maximum [J]. Journal of Dalian Maritime University,2007(04):126-129.

A Hybrid Algorithm for Two-dimensional and Three-stage Exact Packing Optimization Problem

Yifan Sun, Chuang Zhang, Rui Liu, Limin JIA, Yong QIN, Zhipeng WANG*

State Key Lab of Rail Traffic Control and Safety, Beijing Jiaotong University, Beijing, China

zpwang@bjtu.edu.cn

Abstract. In this paper, we study the packing optimization problem of square plates under the conditions that the cutting mode is guillotine cut, the number of cutting stages cannot exceed 3, and the order sequence is not considered. Because it is difficult to determine the constraint conditions in the process of stacking items into stacks, we divide the nonlinear integer programming model into two stages to establish: one stage is the process of stacking items into stacks followed by splicing them into stripes, and the other stage is the process of forming stripes. Finally, a packing optimization algorithm is proposed to solve the problem, which combines an improved bottom-up left-justified algorithm (BL algorithm), greedy algorithm, and iterative sequential value correction algorithm (ISVC algorithm) with a genetic algorithm (GA) as the core. Using this algorithm to solve the data after data preprocessing, the average plate utilization rate can reach 89.29%, which is 10.42% more than the final data without any processing.

1. Introduction

The packing optimization in the personalized customization production mode is essential. The purpose of optimization is to reasonably plan the layout of square parts on the plate so as to reduce the plate waste problem in the blanking process. In this paper, we mainly discuss the packing optimization problem under the following conditions: the number of cutting stages does not exceed 3, that is, items are stacked into stacks, stacks are spliced into stripes, and finally, stripes form plates; the layout scheme is 3H (the length and width are the same) and 3E (one side is equal in length and width) in guillotine cut; the specification of the square plate is $2{,}440 \times 1{,}220\text{mm}^2$, and the quantity is sufficient. And the influence of saw slot width is not considered. The data is from dataset A of question b of *the 19th China Postgraduate Mathematical Modeling Competition "China Optics Valley · Huawei Cup."*

First, we establish a nonlinear integer programming model in two stages. To solve the model, we propose a hybrid algorithm for packing optimization. Suppose the item is directly used as a chromosome, and GA[1~2] is used to perform the entire array operation. The dimension of the action object will be too large, which will affect the time complexity of the solution process and the effective utilization of the plate. So we use different algorithms to arrange items in each process: in the process of stacking items into stacks, we use the improved BL algorithm[3] and stack items in 3H and 3E successively; later, the ISVC algorithm[4] is used to place the stacks in the process of splicing the stacks into stripes; at the same time, the items that cannot be stacked in 3H and 3E mode are directly formed into stacks according to greedy algorithm[5~6]. After the above operations, the chromosomes used to construct plates with GA have changed from item to stack. It dramatically reduces the dimension of the entire arrangement. Not only can the whole arrangement be carried out quickly, but it also can obtain a locally optimal solution that is closer to the global optimal solution.

In actual processing, we found that the order of the data will also affect the effective utilization of the algorithm, so we first preprocessed the original data by reorganizing and sorting, compared the results of the processed data with those of the non-processed data at the end, and determined that the preprocessing in descending order according to the side length will make the effective utilization of the algorithm higher.

2. Data preprocessing

GA is closely related to the data order of chromosomes (chromosomes here refer to item), so the order of data significantly affects the effective utilization of plates. Therefore, we propose two ideas to rearrange the data order:

- Instead of processing the original data set, we directly use the algorithm to solve the problem.
- First, item_length and item_width are exchanged and stitched with the original data. Then, classify them into groups according to the same item_length in the dataset. Then, remove the items with the same ID in the same data type and preferentially remove the same ID between different groups from the group with a small number of items.

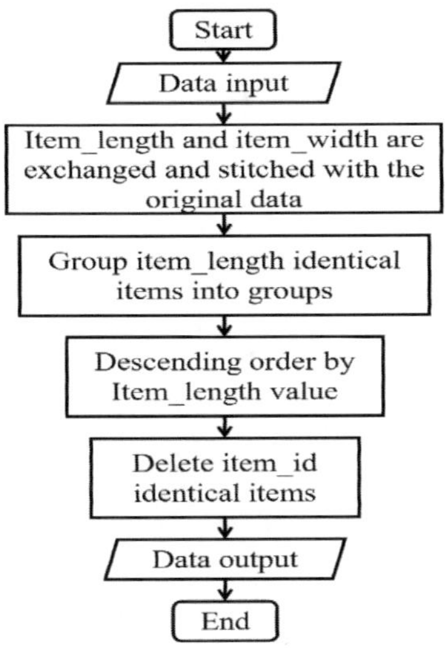

Figure 1. Flow chart of data preprocessing.

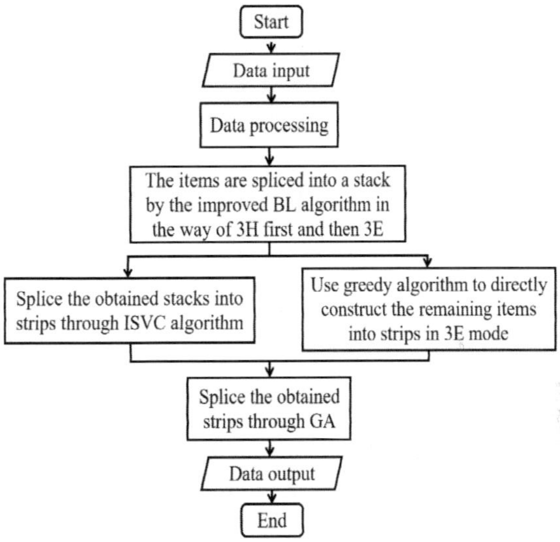

Figure 2. Flow chart of packing optimization algorithm.

Then, we choose two methods to sort groups: one is to sort groups in descending order according to the item_length of different groups, and the other is to sort groups according to the number of items in each group. Through a large number of experiments, we found that the data processed according to the first sorting method had a better effect. Therefore, the grouped data should be sorted in descending order of item_length. The flow chart of data preprocessing is shown in Figure 1.

3. Establishment of the nonlinear integer programming model

The model of the packing optimization problem is actually a nonlinear integer programming model. We constrain the stack and stripe through two stages, respectively.

First, establish the model in the first stage, and introduce two 0-1 decision variables x_{ij} and y_i, where

$$x_{ij} = \begin{cases} 1, & \text{the } j\text{ - th item in group } i \text{ is stacked in the stack,} \\ 0, & \text{the } j\text{ - th item in group } i \text{ is not stacked in the stack.} \end{cases}$$

$$y_i = \begin{cases} 1, & \text{the } i\text{ - th stack is spliced into the stripe,} \\ 0, & \text{the } i\text{ - th stack is not spliced into the stripe.} \end{cases}$$

Suppose that a_{ij} represents the width of the *j*-th item in group *i*, b_i represents the length of group *i*, and the number of groups and items in each group are K and M_i, respectively, so $i \in \{1,2,...,K\}$, $j \in \{1,2,...,M_i\}$.

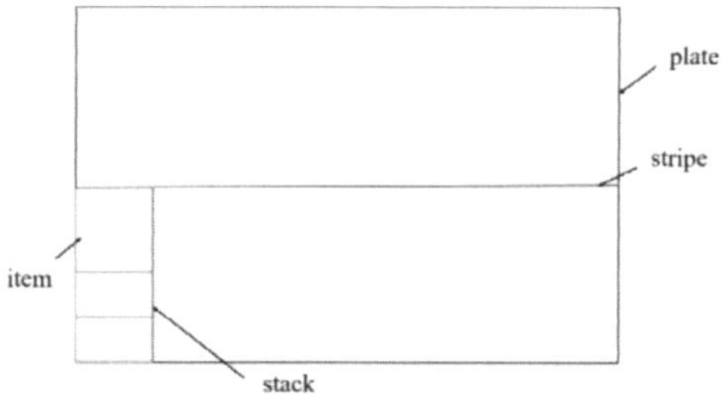

Figure 3. Schematic diagram of stacking items into stacks.

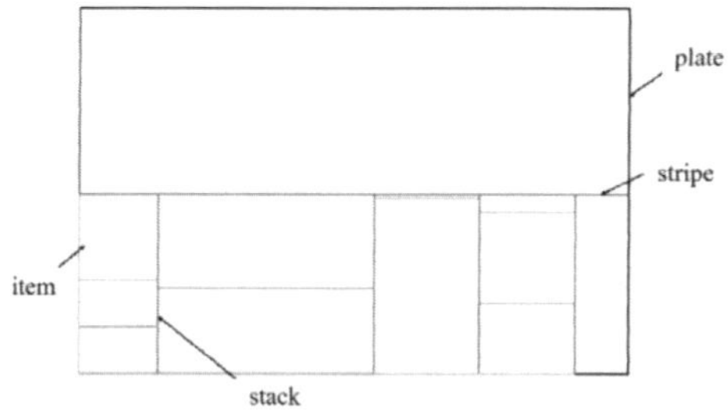

Figure 4. Schematic diagram of stacks spliced into stripes.

The first stage model has two main constraints:

The same item_length in a group should be close to the 2,440mm boundary of the plate, while the item_width should be close to the 1,220mm boundary. The schematic diagram is shown in Figure 3. This completes the operation of stacking stacks from items. The only constraint for stacking stacks is the edge length of 1,220mm, so the first constraint can be expressed as $\sum_{j=1}^{M_i} a_{ij}x_{ij} \leq 1220, i = 1,2,..., K$.

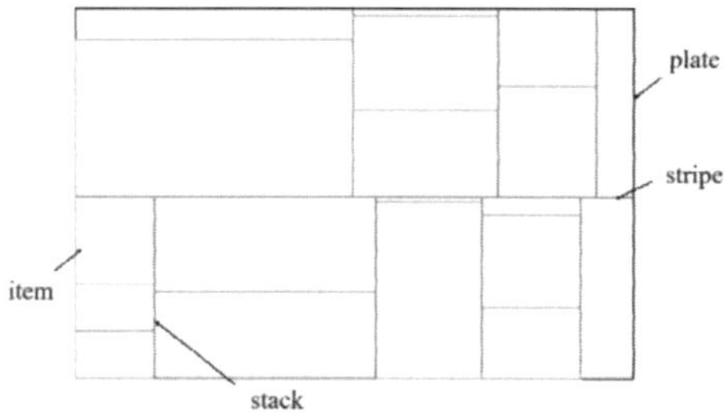

Figure 5. Schematic diagram of stripes forming plates.

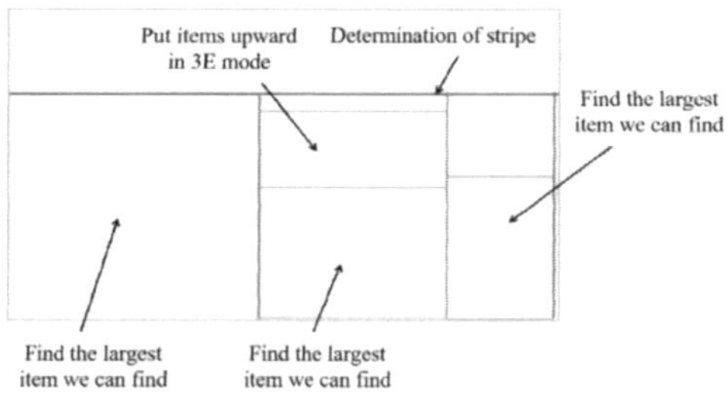

Figure 6. Schematic diagram of splicing the remaining items into stripes.

Then splice the stack from left to right on the boundary of 2,440mm of the plate. The schematic diagram is shown in Figure 4. In this way, the operation of splicing stripes from the stack is completed. Therefore, the constraint condition at this time is to splice as many stacks as possible in the same stripe without exceeding the edge length of the plate. Then the second constraint condition can be expressed as $\sum_{i=1}^{K} b_i y_i \leq 2440$.

The objective function of the first stage is to maximize the utilization area of each stripe, that is $\max\sum_{i=1}^{K} b_i y_i \left(\sum_{j=1}^{M_i} a_{ij} x_{ij} \right)$.

To sum up, the nonlinear integer programming model in the first stage is:

$$\max\sum_{i=1}^{K} b_i y_i \left(\sum_{j=1}^{M_i} a_{ij} x_{ij} \right)$$

$$\text{s.t.} \quad \sum_{i=1}^{K} b_i y_i \le 2440,$$

$$\sum_{j=1}^{M_i} a_{ij} x_{ij} \le 1220, j = 1,2,\cdots,K,$$

$$x_{ij} \in \{0,1\},$$

$$y_i \in \{0,1\}.$$

Next, we will build the model in the second stage, that is, the process of splicing stripes into plates. At this stage, only one 0-1 decision variable z_k needs to be introduced, which means:

$$z_k = \begin{cases} 1, & \text{the } k\text{ - th stripe is spliced into the plate,} \\ 0, & \text{the } k\text{ - th stripe is not spliced into the plate.} \end{cases}$$

Suppose c_k represents the length of the short side of the k-th stripe, and the number of stripes generated in the first stage is N, so $k \in \{1,2,..., N\}$.

The process of splicing stripes into the plate can only be limited by the 1220mm side length, as shown in Figure 5. The constraint condition can be expressed as $\sum_{k=1}^{N} c_k z_k \le 1220$.

The objective function of the second stage is to maximize the sum of the short edge lengths of all stripes that make up a plate, that is $\max\sum_{k=1}^{N} c_k z_k$.

Then the nonlinear integer programming model of the second stage is:

$$\max\sum_{k=1}^{N} c_k z_k$$

$$\text{s.t.} \quad \sum_{k=1}^{N} c_k z_k \le 1220,$$

$$z_k \in \{0,1\}.$$

4. Packing optimization algorithm

Taking Data A1 as an example, if we want to maximize the utilization rate of plates, we need to traverse all 752 data. Therefore, this nonlinear integer programming model is not only computationally expensive and time-consuming but also difficult to find the optimal value.

Therefore, in order to improve the efficiency of the algorithm, we propose a packing optimization algorithm based on GA combined with an improved BL algorithm, greedy algorithm, and ISVC algorithm, following the principle of the minimum number of plates used. The flow chart of this algorithm is shown in Figure 2.

First, preprocess the data and specify the item_ length is length, and item_ width is width.

(a)3H mode. (b)3E mode.

Figure 7. Schematic diagram of two exact layout modes.

Then, to ensure the cutting method of guillotine cut, we propose an improved BL algorithm to stack items into stacks in the way of 3H first and then 3E. The schematic diagram of 3H and 3E modes is shown in Figure 7, where the same number represents that the length and width of two items are equal. The specific idea of the improved BL algorithm is: from the bottom left corner, stack from bottom to top until they exceed the boundary, which constitutes a stack. Whether 3H or 3E, one side of two bottom-up items must have the same length, ensuring that each stack meets the requirements of a guillotine cut.

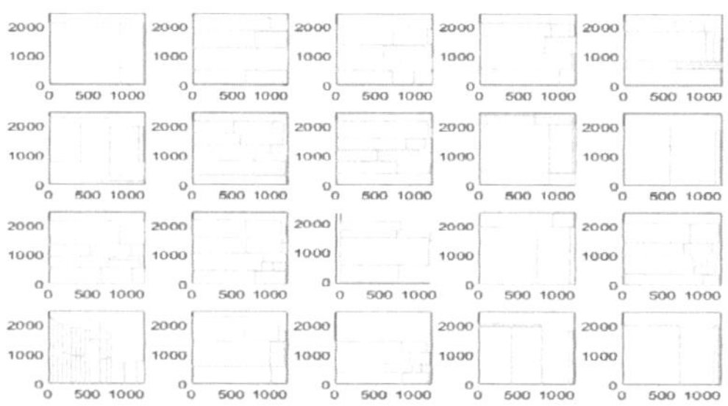

Figure 8. Partial result plots for packing optimization.

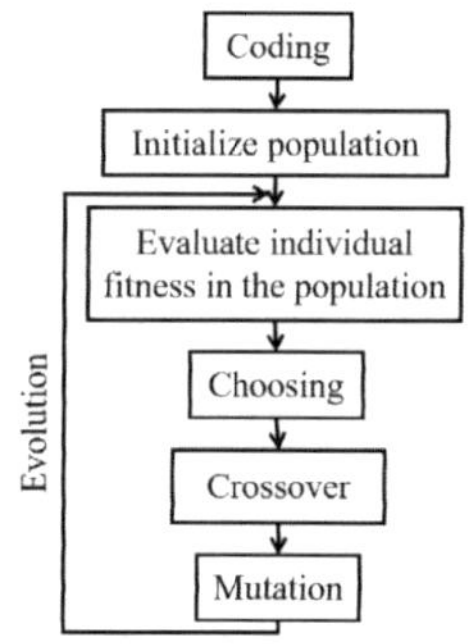

Figure 9. flow chart of GA.

In the process of splicing stacks into stripes, we use the ISVC algorithm. This algorithm is not only beneficial to the stack layout of the current stripe but also retains a certain number of stacks with small areas that are easy to lay out. Then, we can find the packing scheme with the best utilization rate. And the general idea of the ISVC algorithm is: because the stacks with large differences in length will lead to waste when forming a stripe, we form the stripe in the order of stack length from large to small until it exceeds the length of the plate

At the same time, the greedy algorithm is used to directly splice the remaining items into stripes according to the 3E method. The principle of the greedy algorithm is to always consider the item with the largest area first, not the overall optimization, but only the local optimization. The specific splicing method of this process is as follows: first, find the item with the largest area, then treat it as a stack, and then place it on the left of the stripe. We specify that the width of the largest area is the width of the stripe and then place the remaining item with the largest area in the stripe. If they cannot be placed, the stripe will be output. In this process, follow the 3E method to stack the next item into a stack, that is, search for items with equal side lengths that can be placed on the top. If yes, the two items will be stacked into a quasi-stack. If not, the item will be in a separate stack. Then the right side is spliced until the area on the right side is not enough to put down any item so that the operation from item to stripe can be completed. The schematic diagram of this process is shown in Figure 6.

Finally, the stripe is spliced into a plate. After the above two steps, the number of stripe dimensions constituting the plate is far less than 752. Then, the GA is used to fully arrange them to get the preliminary output results, which significantly improves the efficiency of the algorithm and makes it easier to find the optimal solution. We first encode the data, randomly arrange the generated stripe data and express them as chromosomes, and then evaluate the individual fitness, selection, crossover, mutation, and evolution of the stripe population according to the algorithm flow chart in Figure 10. In this paper, the number of evolutions is set to 500.

We use the packing optimization algorithm to solve the preprocessed data and compare the results with those of the data without preprocessing. See Table 1 for the comparison results. And it shows that the packing optimization algorithm is better for grouped and descending data. Therefore, the data should be preprocessed accordingly. Some preliminary renderings of the packing optimization algorithm are shown in Figure 8.

Table 1. Comparison of preprocessing and non-preprocessing results.

	Results without preprocessing		Results with preprocessing	
	Number of plates	Plate utilization rate	Number of plates	Plate utilization rate
Data A1	106	78.81%	95	87.94%
Data A2	107	77.45%	95	87.24%
Data A3	105	79.74%	92	91.01%
Data A4	103	79.47%	90	90.95%

5. Conclusion

This paper describes a packing optimization algorithm based on GA and combined with an improved BL algorithm, ISVC algorithm, and the greedy algorithm. The algorithm can be used to optimize the layout problem under the condition that the guillotine cut, the number of cutting stages does not exceed three times, and the plate material is not considered. The average effective utilization rate can reach 89.29%. But the disadvantage is that it is only applicable to the packing problem under the above conditions, lacking generalization.

References

[1] J Y Feng, Y Liu 2022 Optimization of Evolutionary Genetic Algorithm for Rectangular Nesting Grading for Utilization[J]. *Mechanical Design and Manufacture*:1-6*(in Chinese)*.

[2] W P Tang, K Wang, X Huang 2021 A Hybrid Genetic Algorithm for Two Dimensional Orthogonal Layout of Rectangular Parts [J]. Forging and Stamping Technology, 46(10):106-111 *(in Chinese)*.

[3] H Rao 2019 Solution and Application of Plate Cutting Problem [D]. *Jiangxi University of Finance and Economics (in Chinese)*.

[4] Q L Chen 2016 Research on Optimization Algorithm of Three Stage Layout Scheme for Two Dimensional Cutting Stock Problem [D]. *South China University of Technology (in Chinese)*.

[5] Y M Wu, N Cao, F Y Li, L Li 2022 Research on the optimization algorithm of the plate porous processing path based on greedy algorithm [J]. *Journal of Hefei University of Technology(Natural Science)*,45(06):742-745+759 *(in Chinese)* .

[6] X Chen, F Deng, Q S Yan, Y J Qv 2021 An Improved Texture Layout Method Based on Greedy Algorithm [J]. *Journal of Geomatics*,46(S1):253-256 *(in Chinese)*.

AUTHOR INDEX

Institute of Physics
Dirac House, Temple Back
Bristol BS1 6BE UK

ISSN: 1742-6588
ISBN 978-1-7138-7166-8

9 781713 871668

CURRAN ASSOCIATES INC.
proceedings.com